VGM Opportunities Series

OPPORTUNITIES IN **TELEVISION AND VIDEO CAREERS**

Shonan F. R. Noronha, Ed.D.

Foreword by
Tim Donovan
President
International Television Association

VGM Career Horizons
a division of *NTC Publishing Group*
Lincolnwood, Illinois USA

Cover Photo Credits:
Clockwise from top left: Bell Labs; Jeff Ellis; Preferred Stock; NASA.

Library of Congress Cataloging-in-Publication Data

Noronha, Shonan F. R.
Opportunities in television and video careers / Shonan Noronha.
p. cm. — (VGM opportunities series)
ISBN 0-8442-4090-7 (Hardbound); 0-8442-4091-5 (softbound)
1. Television—Vocational guidance. 2. Video recordings—Production and direction—Vocational guidance. I. Title.
II. Series
PN1992.55.N584
384.55′023′73—dc20 93-20679
CIP

Published by VGM Career Horizons, a division of NTC Publishing Group.

Manufactured in the United States of America.

3 4 5 6 7 8 9 0 VP 9 8 7 6 5 4 3 2 1

ABOUT THE AUTHOR

Writer, producer, educator, and internationally recognized authority on mass media communications, Shonan F. R. Noronha is president of Media Resources, a New York-based consulting firm serving the media, corporations, and educational institutions. Her essays and articles on television and education have been published in journals and industry magazines all over the world.

Noronha was awarded a Doctor of Education degree in Educational Technology by Teachers College, Columbia University in 1979. She also earned Master of Education and Master of Arts degrees from Columbia University. She completed her media studies from the Xavier Institute of Communication Arts, Bombay, after receiving a Bachelor of Arts degree from the University of Bombay, India.

Dr. Noronha was founder and editor-in-chief of several trade publications, among them *Tour & Travel News: The Newspaper for the Leisure Travel Industry* for CMP Publications; and *International Television: The Journal of the International Television Association* (ITVA) for Ziff-Davis Publishing Company. Prior to joining Ziff-Davis, she was technical director of EPIE (Educational Products Information Exchange) Institute.

Noronha's broadcast experience spans two decades and includes production work with PBS (Public Broadcasting Service), New York; Radio Telefis Eireann, Ireland; and All India Radio & Television. She was also TV host of the International Youth Program during the early days of New York's cable television.

She has written and produced training and marketing programs, in a variety of media, for a broad spectrum of corporate and institutional clients. Shonan Noronha has also served on judging panels of domestic and international video festivals.

Dr. Noronha was assistant professor at Iona College, New Rochelle, New York. She was also adjunct associate professor at the Graduate School of Corporate and Political Communications, Fairfield University, Connecticut. She is frequently invited by corporations and universities to lecture on the application of new communications technologies, and her research work is the basis of design decisions made by several communications hardware and software companies.

ACKNOWLEDGMENTS

I did this book like we do every television project—I "networked." My team came through with the assistance I needed. A very special thank you to:

- Lynn Yeazel, western region manager at Panasonic Optical Disk Systems, for his candid narrative of the how, when, and why of his years in the business;
- Jerry Hodges, at Hill Pet Products, for writing the sections on communications satellites and narrowcasting, presented in chapter 1;
- David Eatwell, founder and former president of Houston-based Media Support Group, for succinctly describing what it takes to work in the field of legal video;
- John Rhodes, a New York-based marketing and communications consultant, for providing invaluable information on video hardware sales and marketing;
- Marc Feingold, at Cinram, for his insights on the personal characteristics required for jobs in sales and marketing;
- Billy Bowles, director of VisNet, GTE Service Corporation, for information on corporate video departments; and
- Fred Schmidt, former editor at Knowledge Industry Publications, for assistance in compiling resource material.

For munchies and crunchies, chocolates and flowers; for large doses of humor and much-needed TLC—all credit goes to my network of family and friends. A very special thank you to: Sirikit Noronha, Meena Advani, John Rhodes, Billy Bowles, Jerry Hodges, and Fred Schmidt.

FOREWORD

You'd be hard pressed to name a force that has influenced our society more than television.

It is the medium that occupies more of our time than reading, and more of our children's time than school. It is the medium that brought a war into our living rooms and helped to end it. It is the medium we use to connect ourselves to the world and also the medium we choose to escape it.

Television can soothe or infuriate; it can sharpen the senses or dull them. It can flash sizzling images from anywhere in the world to us in seconds or it can continuously and patiently rerun the distant lives of people named Kramden, Cleaver, and Ricardo.

This now powerful and pervasive medium began quite humbly as little more than radio with pictures. But unlike radio, which is little changed for all of its 80 years, television managed to shake off the shackles of tradition and technology. In the process it changed not only its own character—but nearly everything it touched, as well. Of the influences that unlocked television's potential, few applied exclusively to any single medium. Yet television continued to expand its role—and position in our lives—as the influence of most other forces eroded. Before the end of the 1950s television had established itself as the center-piece of the American living room. And there it stayed.

But while many of us in the 1950s thought that TV's crowning achievement would be color broadcasts of "Bonanza" (viewed through a now curious looking circular screen), there were already forces at work that would shape the future of lives and careers a generation later.

- In remote parts of Oregon and Pennsylvania, so-called community antennas were being raised to pull distant TV signals from the air and feed them by cable to the homes of subscribers. It seemed like little more than Yankee ingenuity at the time, but cable television would grow into a major industry, challenge the traditional dominance of the few broadcast networks in television distribution, and create a hungry new market for video programming.
- Engineers were perfecting a method of capturing video on magnetic tape. The capability to record and play back video was a major breakthrough—hard as it may seem now with today's proliferation of VCRs. Video recording opened the way for the giant corporate television industry and its wealth of video career opportunities.
- A metallic sphere named *Sputnik* was hurtled into space from a secret rocket launching facility in the Soviet Union. What we then called the Space Race was on and the world feared the unknown terrors that would rain down on it from earth orbit. Thirty years later we have nothing to fear but "Mr. Ed" reruns and teleconferences.

Video has continued its advance into the 1990s with improvements not even limited to its own technology; unrestricted by its roots. Computer and laser technologies lent their power to the medium; business and industry embraced it.

Television moved from the living room to the board room. But it didn't *move* as much as it *spread*—expanding its role—pushing ever deeper into our lives. It is a force, for ill or for good, that affects everyone.

And for those of us who make our careers in television and video, it is an exciting force we use daily—and help to shape.

Tim Donovan
President
International Television Association

DEDICATION

To my mother and father,
Sirikit, Antonio, Simkie, and Avertano—
my "private network!"

CONTENTS

CHAPTER 1

NEW TECHNOLOGIES LEADING TO JOB OPPORTUNITIES

The term *broadcast* means "to scatter freely," or in the case of television, to send out signals that can reach a wide audience. Broadcast television has been transmitting to the general public ever since the first commercial signal originated in the United States at the 1939 New York World's Fair. The occasion was a speech by President Franklin D. Roosevelt. Within a few years, the National Broadcasting Company (NBC) and the Columbia Broadcasting System (CBS) began establishing a network of video cables—called *coaxial cables*— that would allow them to route signals from their New York facilities to all their affiliate television stations across the country. They provided television coverage of primarily news and sporting events that were in turn broadcast by affiliate stations to the local population. For years, this network of coaxial cables served as the only link between the network and its affiliates.

SATELLITES

Beginning in the mid-1960s and increasing throughout the 1970s and early 1980s, communications satellites were launched into orbit around the earth. Most television networks now use satellites or digital fiber optic networks to distribute their signals to affiliate stations for retransmission to a local population. Satellites have become so popular for distribution of information that an entire new industry has developed around their use. The concept that any number of satellite dishes can be installed over a wide geographic area to receive live television information has intrigued more people than just broadcasters. Many companies are finding that satellites offer solutions to problems associated with the dissemination of information.

An understanding of the basics of satellite technology is helpful in realizing the role it plays in the communications industry. It's really quite simple. There is a circular orbit in space, about 23,000 miles above the earth, that is called the *geosynchronous belt*. This belt lies directly above the earth's equator. A satellite that is launched into orbit within this belt can be orbited at the same speed at which the earth rotates, so that the satellite appears stationary from the earth. The 23,000-mile distance is the point at which there is a balance between the gravity that tries to pull the satellite back to earth and the centrifugal force that tries to cast it further into space. The geosynchronous belt is the only location in space that allows a satellite to maintain precise relative positioning to the earth.

The satellite itself is nothing more than a relay station. It receives signals transmitted to the satellite from the earth on a preassigned *frequency* (channel). It then retransmits the signal back to the earth on a different frequency. The signal transmitted back to the earth usually covers a wide area. Most U.S. satellites cover all of the continental United States and parts of Canada and Mexico.

There are three *bands,* or frequency groups, that have been approved by the Federal Communications Commission (FCC) for use by the public sector. The older satellites are in the C-band that operates at the lowest-frequency ranges. Most C-band satellites have 24 video channels and generally operate with a low power output, sometimes as low as five watts. Because of this low output, a large field-array antenna or dish antenna is needed to capture and reflect the signals back to a receptor.

The second most popular is the Ku band. This band operates on the highest frequency and generally transmits with a higher power output than do C-band satellites. Many of the Ku satellites operate with power outputs of 40 watts and more. Because of the higher output levels and characteristics related to the higher frequency, antenna arrays Ku satellites can be smaller than the arrays required for C-band. One drawback of the Ku band is that it is more sensitive than C-band satellite signals are to *atmospheric attenuation,* that is, fog and heavy rain that can affect quality of the picture.

The third band is the Ka band. The Ka band has been designated for use by high-powered *direct broadcast satellites* (DBS) that will operate services targeted directly to the home television viewer. DBS technology has been slow in gaining popularity, but it holds great potential. Its main advantage over the other two frequency bands is that it transmits with as much as 200 watts of power, meaning that an inexpensive, smaller dish or array can be used to receive the signal. Some services specify dishes as small as 2.5 feet in diameter.

All of this activity in the development of satellite technologies and the recognized power of broadcast television have heightened awareness of the potential power that these advances hold in solving problems. This awareness has led to entire new industries and opportunities in the video production business.

In the 1960s and 1970s, broadcast television became the most popular method of distributing the news and providing entertainment for the American public. But in recent times, television has proven important to our society in other ways. Corporations today are turning to television to solve problems in communications and training.

The use of television in corporate environments has been refined to the point that much of the material being produced today compares favorably in quality with the programming being developed for broadcast television. Corporate television has proven to be one of the most useful training and information tools available to companies and institutions to date. Almost all of the larger corporations in the United States have a video presence within their companies to some degree, and many smaller firms are relying on video to solve some of their training and communications problems. But the use of television by industry has grown beyond the confines of the videocassette player.

NARROWCASTING

Today, industry leaders are turning to satellite distribution for dissemination of their visual material. Acceptance of the use of satellites in industry has led to the development of an entire new industry in communications.

This new industry has been named *narrowcasting*. It takes the best ideas of broadcast television and applies them directly to corporate television. Narrowcasting transmits a signal to a specifically defined audience instead of to the general public. The audience is targeted, much like advertising agencies target their commercials to particular age, sex, or ethnic groups. An entire industry, such as the banking industry or the restaurant industry, may be targeted for narrowcasting service.

This trend has accelerated with the proliferation of multichannel cable TV operations. With 30, 50, or even 100 channels available, many cable operators have found it desirable to carry satellite channels aimed at a very narrow audience (e.g. doctors, women, ethnic and religious groups, etc.)

The use of a satellite for narrowcasting works only when the potential audience is large and diversely scattered. The cost of transmitting over a satellite can be high, and until recently, the cost of satellite receiving equipment was prohibitive except in the most crucial of situations. But several events in the mid-1980s brought the overall costs down significantly and have made narrowcasting feasible for many industries.

First, many communications-related companies anticipated that demand for satellite channels would be strong in the 1980s. However, this demand did not materialize as they had predicted. As a result, a surplus of satellite time was available, and satellite owners became willing to make deals for broadcast time. Another event was the development of excess production capacity in the receiver area. Many manufacturers had anticipated that a large home market for satellite dishes would develop. This trend began in the early 1980s but dwindled as more and more satellite services began encrypting signals.

Entertainment services such as Home Box Office (HBO) initially distributed their signals to subscribers over satellites without any protection of their signal. In other words, anyone who wanted to receive the signal simply needed to install a receiving dish to pick up these "premium" channels without paying for the service. HBO and the others gambled that the high cost of receiving dishes would be enough deterrent to keep most people who weren't HBO subscribers from receiving their movies. However, the demand for dishes was great, and manufacturing capacity expanded to handle the anticipated volumes. By the mid-1980s, HBO and others began to scramble the signals, which in turn diminished the demand for dishes. Thus, a surplus capacity of receivers resulted and prices fell.

NARROWCASTING BUSINESS TELEVISION

Today, receiving systems can be purchased for a fraction of their original price, and in some instances the receiving systems cost no more than a medium-priced videotape recorder. This lower pricing structure in part created an economic environment that helped fuel the development of narrowcasting. Receiving systems could be purchased and installed relatively inexpensively.

Revenues for narrowcasting are generated in a variety of ways. One popular method is by selling subscriptions. The operator of the service might scramble the signal and offer to provide receiving equipment and decoder boxes to those businesses that could benefit from the programming. With a subscriber base of only 1,000, a narrowcasting service can be profitable.

Another method is by offering the use of the network to the industry it serves for a price. By leasing time on the network to companies that have multiple field locations, the services have been known to generate more revenues than through subscriptions.

A third method of generating revenues is by running advertising on the network for a fee, similar to standard broadcasting. Obviously, the selection of the industry that a narrowcasting service chooses to serve can dramatically affect its ability to generate revenues.

Most narrowcasting services select industries that have certain generic needs. They then develop a curriculum to address the problems and establish a "program day." At some time during each business day, narrowcasting services transmit news and information about the industry they serve, as well as offer specific training. Based on the appropriateness and quality of the programming, companies within that industry may choose to subscribe to the programming.

Subscribers must have a satellite receiving system and, in most cases, a signal decoding box that will allow them to receive the signal. For a nominal fee, the narrowcasting service offers up-to-the-minute industry news and effective video training, usually of a higher quality than could be provided by the company itself.

ONE APPLICATION OF NARROWCASTING

The restaurant industry is a good example to use in explaining how a narrowcasting service can work. This industry has a relatively high employee turnover rate and a moderate-to-low payscale. Most restaurant managers will agree that success is based on the quality of both the product and the service. This is the toughest challenge faced by upper management of many restaurant chains and is the reason that they have difficulty maintaining consistency from store to store.

Since the restaurant business is a relatively low-profit-margin industry, it is imperative that quality be maintained at minimal expense. Most chain restaurants cannot afford to send all of their employees to training schools to learn how to deal with people, how to properly cook the food, how to maintain sanitary conditions, and other procedures. Some have turned to producing videotape presentations. But even with this solution, the cost of producing the material is high. The solution could be an industrywide service that produces generic materials to provide basic training to a variety of restaurants and a diverse customer base.

This is exactly the approach that narrowcasting takes. By signing up subscribers within a particular industry, the cost of a single production can be spread over a larger number of users. Distribution of the programming is offered daily and is easily accessed by simply turning on the satellite receiver at a particular time of day.

Programming must be of a quality equivalent to regular broadcast television, and in some ways, the production of this material takes more creativity than normal entertainment programming on regular broadcast television. Care must be taken to provide training and information that is unbiased and serves the entire customer base, not just a segment of that base.

Narrowcasting offers other benefits besides training and industry news dissemination. In the example of the restaurant business, many chains are spread across the country. When a chain signs up with a narrowcasting service, the satellite receiving systems that are required for standard reception can be accessed by management for directing specific programming directly to their stores. The scrambling systems utilized in many of these networks are capable of selectively authorizing reception, based on a secret code stored inside the descrambler.

For example, a chain of steak houses might want to introduce a new dish to its chefs and show them how to properly prepare it. In such an instance, the steak house chain would not want anyone else to see the specifics of how the dish is prepared. The narrowcasting service would simply turn off all decoders except those at the designated locations, and the signal would be seen and heard only by a select group.

Access to a national base, such as described above, can provide management with a critical advantage, can save many trips that otherwise would be required for management and training meetings, and can provide information in a more time-efficient manner than would traveling to a common location for a meeting. In many instances, the benefits offered by narrowcasting services can more than pay for themselves in time and travel for a number of industries. The future of this industry is bright and limitless.

Narrowcasting as a Career

For individuals entering the job market, this emerging industry holds much of the excitement experienced in normal broadcast television. Each narrowcasting network is essentially a broadcast television station. Staffing is similar to that of a commercial television station, but since the environment is corporate television, it can be far more appealing. All of the elements and excitement of broadcast television are there: the stories ready by "air time," production of live programs, travel, and creative production. Since it is an emerging industry, narrowcasting for corporate television offers opportunities for an individual to advance that are far more promising than at broadcast television stations. Many of the same job categories exist—camera operation, producer, editor, sales personnel—all are a part of the complete narrowcasting service. In addition, technicians are required to install and maintain receiving systems. Engineers are also needed, since each network broadcasts over a satellite *uplink* (transmitter).

Many corporate television producers have chosen to stay in corporate production rather than move over into commercial television broadcasting. They find the challenge they desire and have a greater latitude for creativity in the corporate environment.

Narrowcasting offers many of the benefits of corporate television production along with the element of excitement that is experienced in broadcast television. It is a new career path that should be explored by newcomers to the industry as well as those seeking a midcareer change.

"TV On-Demand" or Interactive Television

An emerging technology that will increase the need for targeted programming is an interactive "broadcast" medium sometimes referred to as "TV On-Demand." An outgrowth of developments in cable television and telecommunications technologies, interest in "TV On-Demand" has been fueled by recent court decisions allowing phone companies (RBOCs) to enter the electronic information delivery business.

This technology permits users to select from dozens or hundreds of shows and services delivered to their TV and/or home computer via digital phone lines or coaxial cables.

The need for additional programming "product" for Interactive TV should produce a strongly increased demand for creative talent in this area in the near future.

VIDEOCONFERENCING AND BUSINESS TELEVISION (BTV)

The availability of low-cost Ku band technology has enabled many organizations to create private television networks. Most organizations use these networks to conduct long-distance conferences known as *Business Television or satellite videoconferencing*.

In the early 1980s, teleconferencing networks were temporarily created with portable uplinks and downlinks and permanently installed antennas in public facilities such as hotels or in television studios. At that time, the costs for setting up such a meeting were high, so teleconferences were held as special events.

Since the mid-1980s, the demand for these meetings, indeed for long distance training and presentations of all kinds, has increased dramatically; and the necessary technology has become more flexible and less expensive to employ.

Today, most point-to-multi-point conferences are broadcast via satellite, and are called "Business Television," or BTV. They involve a one-way video and audio transmission, with questions or tabulated responses sometimes being relayed back to the originating studio via telephone. Programming can be live or taped, and is usually of high quality, employing the best available teleproduction equipment and talent. This medium is ideal for "spreading the word" quickly to a large audience, in announcing a new sales program, or disseminating new procedures and policies to an entire organization.

Other conferences involve fully interactive two-way video and audio transmissions, and are generally referred to as "Videoconferences." Two-way, point-to-point videoconferencing is used to facilitate meetings between small groups of people in two, or on occasion, as many as eight, remote locations; and is usually conducted via digital land lines.

Many companies use videoconferencing to reduce travel costs. For example, the cost of high quality video transmission from San Francisco to New York can be less than $300 per hour, while a round-trip airline ticket during peak business hours can cost over $1,000 per person. A company using videoconferencing in lieu of sending a team of six people to a meeting saves considerably on time and travel costs.

Lower quality "compressed" video transmission, similar to that transmitted from the Space Shuttle (but quite adequate for information "shirt sleeve" meetings) can cost less than $50 an hour, coast to coast. A major reason for the rapid growth of two-way interactive videoconferencing is the great increase in productivity it provides. Reducing the time required to make decisions, and to respond effectively to competitive pressures, produces a significant edge, the value of which has not been lost on corporate management. Combine this with the greatly reduced costs of transmission and hardware (over 75% over the last five years), and it is no surprise that the sales (and use) of two-way videoconferencing systems has increased so rapidly.

Companies have found that videoconferences are better structured and the agendas more carefully prepared and adhered to than at traditional meetings. In addition, videoconferences are usually shorter than traditional "in person" meetings.

Today, business television conferences at some companies involve hundreds of people at multiple locations in meetings that often last up to six hours. Tektronix, Inc., for example, conducted a six-hour technical training session via a teleconference. The one-way video, two-way audio session introduced two new Tektronix instruments to its 200 sales engineers. Hospital Satellite Network produced and transmitted "Who Cares for the Poor," a 90-minute videoconference in which 200 hospital employees examined the position of the federal government on this issue.

Universities have begun using videoconferencing to conduct employment interviews of students. Georgia Institute of Technology is reportedly the first university to use it for recruiting. Georgia Tech students have been interviewed by recruiters from General Electric, Harris Corporation and Boeing Company. Boeing also has used the system to interview students from the University of Florida at Gainesville and the University of Tennessee at Knoxville. The video interviews are said to take the place of the student traveling to the company for the second or intermediate interview. The video conference, while making it possible to have a face-to-face interview, reduces the massive expenditures that companies incur in flying large numbers of college students to corporate locations for intermediate interviews, following the initial interview on campus.

Spacenet, GTE's videoconferencing network, has identified the following advantages of videoconferencing for business applications:

- More timely decision-making: meetings are arranged quickly and easily, whenever the need arises; information gathering is fast and easy; key people at remote sites are readily available.
- More effective meetings: all the materials in your office are readily available for your meeting; staff members can easily attend or be called in on short notice.
- Improved communications between remote locations: meetings can be held more frequently because they are more easily arranged.
- Expanded range of control: managers can monitor projects at remote locations.
- Greater impact: video images allow nonverbal communication as in a meeting conducted in person.

According to the International Teleconference Association (ITCA) more than 100 Business Television Networks are in operation in the U.S. alone, with over 30,000 receive sites (or downlinks) in the U.S. There are more than 5,000 two-way videoconferencing rooms currently in use by corporate and institutional organizations.

While the number of new BTV networks is not growing rapidly, the "viewership" and the number of downlink sites is expanding at a healthy pace, driven by the need for more rapid and effective corporate communications. Emerging technologies, such as digital compression (Spectrum Saver), and desktop delivery (PC-TV), are also fueling strong growth in this area.

The explosive growth of two-way videoconferencing has not produced a corresponding increase in demand for video production expertise, probably due to its emphasis on less formal meetings and its use of lower quality (highly compressed) video. The biggest needs in this area will be for "meeting production" skills, including multimedia presentation production expertise, and the skills that relate to planning and the facilitation of meetings.

INTERACTIVE VIDEODISC

Interactive video is the term for the reciprocal action called for on the part of the user of such video programs. Interactive video uses several technologies: videotape, videodisc, and computer. This marriage of technologies offers exponential capabilities.

Despite the erratic history of the videodisc industry, current technological developments such as laser videodiscs, erasable optical discs, and still-frame audio for laser discs are impressive and useful. While the single largest application of videodisc technology has been training, today it is increasingly being used at point-of-sale, for archival storage, and in arcade games.

Interactive Video and Multimedia

Recent developments in the multimedia area such as Apple's Quicktime™, Ultimedia™ from IBM, and the widespread availability of CD-ROM hardware and software, have greatly increased the ability of "ordinary" personal computers to utilize video and audio. This growth has spread from games, instructional programs, and point-of-sales (POS) into mainstream business applications such as word processing, telecommunications and presentations programs. The rapidly expanding multimedia production business requires individuals with an amalgam of skills including video and audio production/post-production expertise; and computer skills such as flowcharting, "C" language programming, and user interface design.

Laser videodiscs provide the information density more commonly associated with the print than the video medium. Each videodisc can hold 54,000 frames of information. For example, the Library of Congress has stored, on videodisc, volumes of printed material which would otherwise have occupied several thousand square feet of space.

The Doyle Dane Bernbach advertising agency uses videodiscs for storage of commercials. Prior to videodiscs, the company stored commercials on hundreds of videotapes. Now, the company is able to store innumerable commercials on a single disc. Since each disc has two separate audio tracks, the sound track of television commercials is put on one track, while radio ads are stored on the second track.

An interactive videodisc exhibit sponsored jointly by *Vogue* magazine and the Cadillac Motor Division of General Motors targeted a growing market of young women professionals. It aimed to change their perceptions of the Cadillac line and get them behind the wheel of a Cadillac for a test drive. The exhibit, entitled "Expressions," was shown in shopping malls in California. It featured a videodisc kiosk with an infrared touchscreen controller.

To create videodisc programs, there is a growing need for not only people who possess conventional video production skills, but also for writers, producers, and computer programmers who can do disc preproduction work such as flow-charting. Technological advancements have opened whole new job markets to people trained in various specialized areas of video production.

COMPUTER GRAPHICS

Computer graphics is one of the fastest-growing segments of the visual communications industry. It has enhanced the presentation quality of business information tremendously.

Today's computerized imaging devices are a great boon. They allow people to create, manipulate, store, and retrieve images with artistic flair, flexibility, and speed. Digital video effects (DVE), electronic paint, and animation systems offer producers astounding special effects to give impact to their message.

Computer graphics engineers can select a palette from literally 16 million colors. Some systems allow the producer to select a pattern or texture and then insert it into a specified area. Color mixing is a universal feature of these systems, as are freehand sketching and painting.

Once limited to switchers, image manipulation can now be performed on several specialized units. Wipes, mattes, simple animation, and the relocation of still pictures across the screen—all this is now within the grasp of even low-budget producers. Millions and millions of dollars are being spent on computerized images today with corporate and "cable" oriented production accounting for a large share of this growing market.

The dramatic reductions in the cost of the technology has produced in this area, more so than in any other "video" discipline, the expectation for high production values and the skills to produce them. For example, animations that ten years ago were produced on IBM mainframes in one week, can be done today on a Mac or Amiga in a few hours; and title graphics that required a $50,000 character generator in 1985, pale next to those generated on a $5,000 PC in 1993. Corporate and cable clients have come to expect that these dramatic visual effects will be part of all but the most "bare bones" productions; and this expectation has increased the demand for more creative teleproduction professionals.

What all this means to a person considering a career in the field of television is that there are many more job opportunities available today, especially in the new areas of specialization: narrowcasting, business television, computer graphics, and multimedia.

CHAPTER 2

BROADCAST TELEVISION

Television is the youngest, yet most powerful, medium of mass communication today. Its ability to grab the viewer, both emotionally and intellectually, has made it emerge as the dominant mass media communications form of our time. It is, in fact, the all-pervasive medium—the instant informer, effective teacher, primary persuader, and great entertainer.

Television has dramatically changed the nature of news. It has given us the ability to deliver news as it happens—instantly, accurately, and vividly. Television as a teacher has revolutionized the traditional American concept of education. It can make learning even more interactive and interesting. The persuasive power of television messages commands billions of advertising dollars annually. Television encourages consumers to buy, politicians to mold public opinion, and preachers to spread "the word." As an entertainer, television brings the spectacular right into the living room. Indeed, television seems to have tremendous influence upon our society in an infinite variety of ways.

Rapid technological advances in the television industry have widened the scope of its application to all sectors of society. The proliferation of new cable channels has led to a cable penetration of 45 percent of the over 85 million TV households in the United States. This widespread use of television has led to a multibillion-dollar industry, generating employment opportunities in the programming, production, administrative, marketing, and technical areas.

FEDERAL COMMUNICATIONS COMMISSION

In 1934, the federal government of the United States established the Federal Communications Commission (FCC) as an independent federal agency to regulate radio and television stations. The FCC is responsible for: 1) the allocation of space in the frequency spectrum; 2) the assignment of stations, with specific frequency and power, within the allocated bands; and 3) the regulation of existing stations, to ensure that they operate in compliance with FCC rules and technical provisions.

Television stations are permitted to broadcast only on assigned channels. Channels 2 to 13 are designated to stations broadcasting in the VHF (very high frequency) bands. All stations assigned channel numbers above 13 operate in the UHF (ultra high frequency) band. Some 2,000 channels are in operation in 1,300 communities.

COMMERCIAL BROADCASTING

Commercial Networks

Nearly two-thirds of the channel allocations are to commercial stations. Hence, the majority of jobs in the broadcast industry are available at commercial stations.

The largest commercial TV networks are Columbia Broadcasting Services (CBS) Corporation, National Broadcasting Corporation (NBC), and American Broadcasting Corporation (ABC).

These three networks have corporate headquarters in New York City and studio facilities there as well as in Los Angeles, Chicago, London, and other major cities worldwide. Some commercial television stations are associated with networks and receive their programs via microwave or satellite transmissions. Such stations are known as *network affiliates*.

Independent Stations

About 21 percent of all commercial television stations are not affiliated with a network. Such stations are commonly referred to as independent TV stations. According to the Independent Television Stations Association (INTV), there were 422 independent stations on the air in 1993.

Independents use a variety of sources for programming. While it is not uncommon for an independent station to rely heavily on the use of motion pictures and reruns of programs previously shown on the networks, many of them also generate local news and other original programming. Several nationally known broadcasters began their careers at independent stations.

In 1991, commercial television stations employed 65,598 people, of which 20,095 were classified as professionals, according to the FCC. Networks generate original programming and therefore provide a great number of job opportunities.

NONCOMMERCIAL BROADCASTING

More than one-third of the licenses issued and channel assignments made by the FCC are for noncommercial use. In 1952, the FCC made channel assignments to 242 communities for noncommercial educational TV stations. The first noncommercial stations began broadcasting in 1953.

Public broadcasting stations are operated by nonprofit institutions such as local community groups, universities, and religious institutions. The stations are operated as nonprofit corporations, hence they do not sell time for commercials. Public broadcasting stations are expected to provide programming on topics that usually would not be supported by commercial stations.

Public Broadcasting Service

In 1967 Congress passed the Public Broadcasting Act, under which federal funds were allocated for the support of public broadcasting. The Corporation for Public Broadcasting (CPB), was set up as a nonprofit corporation, responsible for receiving, distributing, and general administration of federal funds for the entire system. The Public Broadcasting Service (PBS) was established to manage the production and distribution of programs and connections among local stations within the system. PBS is the largest noncommercial network in the United States and has been referred to as the "fourth network."

The role of PBS in relation to local stations is similar to that of the commercial networks to their affiliates. However, PBS affiliates have more of an independent operating status than that of commercial network affiliates. PBS affiliates are active in producing and distributing programs nationwide. In 1993 PBS was providing national programming and research to 347 affiliated stations. PBS stations usually broadcast educational television programs as well as some cultural programs covering subjects such as dance and drama.

In addition to federal grants, funds to cover operating expenses for public broadcasting come from agencies such as state governments, universities, religious organizations, and corporations. Due to lack of adequate funding, PBS growth has been stifled. Many public stations allocate air time for pledge drives, when they solicit financial support from the audience by urging viewers to call the station and pledge money to support it. In addition, fund-raising efforts at several PBS stations include soliciting business sponsors to underwrite the cost of a program or special series and renting out the station's production facilities and services for teleconferences.

Although the number of public broadcasting stations is not expected to increase significantly over the next few years, the employment opportunities in production and technical areas will increase in accordance with increased financial support. In 1992, public television stations employed 10,912 people, of which 46.3 percent were women and 18.5 percent were minorities, according to the Corporation for Public Broadcasting.

Instructional Television Fixed Services (ITFS)

In 1963, the FCC established the ITFS as a form of supplementary education. In 1983, there were 83 ITFS licensees operating 644 channels, according to the Network for Instructional Television (NITV). Also in 1983, the FCC issued new regulations permitting educational ITFS systems to lease their unused channels to commercial firms. According to a study conducted by the Corporation for Public Broadcasting, ITFS audiences are primarily students from kindergarten through grade 12 and adults seeking further education or training in industry.

Television Stations on the Air as of November 30, 1992

	Commercial	*Educational*	*Low Power*	*Total*
VHF	558	124	469	1,151
UHF	588	239	842	1,669
TOTAL	1,146	363	1,311	2,820

Source: FCC

CABLE TELEVISION

Several communities are outside the broadcast range of networks, independent TV, and public broadcasting stations. Consequently, they receive poor (if any) reception of regular broadcasts. Cable television (CATV) originated as a means of bringing better reception of regularly broadcast programs to communities on the fringe areas of transmission service.

The two earliest cable television installations date back to 1948, in Lansford and Mahanoy City, Pennsylvania. However, it was not until 1983 that cable television began to expand significantly.

Cable television is subscriber-supported. Subscribers pay a fee to be able to watch programs on cable television. Usually, subscribers pay a basic cable service fee to receive better reception of programs from local stations and from a few stations outside the local area.

Subscribers have the option of paying additional fees to receive extra, premium pay-cable services. Usually, there is a charge per channel. Pay-cable networks such as HBO, Showtime and Cinemax carry primarily first-run Hollywood movies and other special entertainment programs carried exclusively on their network.

Some cable television services receive additional support by selling air time for advertising. Cable satellite networks which are ad-supported include CNN and CNN2, both 24-hour news channels; ESPN, a 24-hour service covering sports events and sports news; Cable Health Network; the Weather Channel; Nickelodeon, a service featuring children's programs; and Modern Satellite Network.

Most cable operators find it cheaper and easier to buy programming than to produce it themselves. Hence, the majority of job opportunities in the cable industry are in the engineering, sales, and customer service areas. However, FCC regulations require CATV stations that serve over 3,500 subscribers to originate a certain amount of local programming. By 1992, cable employed 108,280 people nationally, according to the FCC.

LOW-POWER TELEVISION

In 1982, the FCC started granting permits for the construction of low-power television (LPTV) stations. Under FCC rules, LPTV stations are limited to 10 watts VHF and 1,000 watts UHF. These low-power stations cover a distance of less than 20 miles. They are designed to serve specific audiences such as small rural communities or ethnic groups in large cities. According to the FCC, low-power stations have a *secondary status,* which means that they are expected to avoid transmission interference with full-power stations that command *primary status.*

Currently, some low-power television stations are operated as nonprofit corporations; others that sell commercial time do not have nonprofit status. A few LPTV stations scramble their signals and operate on a subscription basis. That is, they charge viewers a subscription fee, in much the same way pay-cable stations operate. By 1985, there were an estimated 344 LPTV stations on the air, including 196 stations in Alaska, according to Kampas/Biel and Associates. By 1992, that number increased dramatically to a total of 1,311 LPTV stations.

CHAPTER 3

TELEVISION AND VIDEO IN NONBROADCAST ORGANIZATIONS

One of the rapidly growing sectors of the television industry is the corporate/institutional area. Corporations as well as non-profit institutions use video extensively for employee communications, staff training, and the marketing and promotion of products and services. The majority of corporate and institutional video programs are not broadcast. This sector of the industry has been called nonbroadcast television, industrial television, corporate television, and private television. Names used more frequently today include organizational TV, corporate TV, business TV, and professional video.

The field of organizational television embraces all departments that use video for professional communications. These include government agencies and corporations as well as not-for-profit institutions such as universities, hospitals, and museums. While the majority of programs produced by these organizations are not broadcast, some large corporations do "transmit" and therefore "broadcast" some of their programs. Several large companies are using sophisticated video technologies such as teleconferencing and interactive videodiscs to meet their complex communications needs.

The foremost industry association in the organizational TV area is the International Television Association (ITVA). Its members represent corporations; health, medical, and educational institutions; video production and postproduction facilities; and manufacturers, dealers, and distributors of television equipment. Many of these companies have full-fledged video departments staffed by managers, producers, directors, writers, designers, engineers, and technicians.

HOW CORPORATIONS ARE USING VIDEO

ITVA members produce video for different industries. Among them are:

Aerospace	*Government/Military*
Computer	*Education*
Financial Service	*Emergency Services*
Hospitality	*Health Care*
Legal	*Nonprofit Organizations*
Manufacturing	*Religious*
Petroleum and Gas	*Public Utilities*
Transportation	*Media Services*

The development of the half-inch cassette recorder has led to the expansion of corporate TV departments into corporate video networks. Half-inch video as a distribution medium has made it possible for corporations to communicate from headquarters with their branch offices several thousand miles away.

The past decade has seen a growing number of corporations making an investment in video networks. The most dramatic expansion has taken place in the area of video program distribution locations. While some corporations have video networks reaching 25 locations, larger networks distribute corporate programs to more than 1,000 locations.

For example, Burger King Corporation, the Miami, Florida-based hamburger giant, uses video in over 1,100 company-owned and franchised restaurants to train employees and assure adherence to its stringent operational standards. In the fast food industry, where high employee turnover is a dominant trend, the company faces the constant challenge of communicating proper operating procedures to more than 100,000 crew members on duty at any given time. Burger King is finding video a very effective training tool for restaurant personnel.

Video represents a major investment in marketing communication made by Casual Corner, one of the larger women's clothing chains in the United States, with headquarters in Enfield, Connecticut. The company utilizes video for sales training and point-of-sale presentations in its 750 stores nationwide. Video is helping the staff at retail stores build volume sales. Some programs teach the staff how to better sell the clothing, while others enhance the merchandise presentations with consumer education. The VHS video cassette player and 19-inch color video monitor add to the excitement of product demonstration for customer viewing in the store.

Although skills training and management communications have ranked high in corporate applications of video, the area of rapid growth is employee communications. Several corporations are expanding their video networks so as to reach as many employees as possible in the shortest period of time. Leading corporate communications departments are giving employee news, employee information, employee orientation and employee benefits programs top priority. For example, Ford Motor Company in Dearborn, Michigan, utilizes a sophisticated video communications system exclusively for employee relations. The company is said to have the capability to reach more than two-thirds of its approximately 250,000 North American employees within a half-hour, which allows the company to adjust its policies and procedures in a very timely manner.

According to a study conducted by D/J Brush Associates, in 1985 the top ranked use of video was in skill training. Seventy-five percent of those responding to the survey reported producing programs in that category. The study identified the following video applications:

Skill Training
Employee Orientation
Employee Information
Employee Benefits
Job Training
Management Communications
Community Relations
Employee News Programs
Annual Reports and Meetings
Professional Upgrading
Point-of-Sale
Safety and Health
Supervisory Training
Sales Training
Management Development
Product Demonstration
Sales Meetings
Economic Information and Education
Government and Labor Relations
Security Analyst Presentations

Profiles of Corporate Video Departments

Phillips Petroleum Company in Bartlesville, Oklahoma, started an in-house video facility in 1976, primarily to produce programs for training and management communications. Today, the department also produces programs for employee communications and product promotions. According to Scott Carlberg, supervisor-producer, the department produces 60 to 70 programs annually, including a biweekly employee news program entitled, "Inside Phillips." The video department consists of seven full-time staff which includes a supervisor-producer, two producers, one chief engineer, one assistant engineer, one production assistant, and one librarian.

Amway Corporation in Ada, Michigan, started its video department in 1983 with a staff of 15 people. Today, its staff of 45 includes 12 producer-directors as well as videographers, editors, engineers, and other production staff. According to Ron Brown, audiovisual manager, the department produces approximately 80 programs a year. The majority of programs are for training, customer information, and sales meetings.

GTE has a long history of leadership in the corporate video arena. Over the past two decades, GTE has established production and broadcast centers in many of its strategic business units. In 1989, GTE Corporation consolidated these resources into a nationwide network, under the registered name, GTE VisNet ®. The name VisNet stands for Visual Communications Network.

The GTE VisNet team of communications professionals provide creative and technical expertise in all areas of video and television production, engineering and system design, multimedia and other interactive communications technologies within and outside GTE. They operate from nine locations with Stamford, Connecticut and Dallas, Texas as their major production hubs.

VisNet is a full service communications provider with product lines as follows: Communications Consulting, Business Television (BTV), Interactive Distance Learning, Video Production and Distribution, Graphics Services, Advanced Communications Services: Multimedia Design and Production, and Knowledge Networking.

In the area of Business Television, VisNet provides network management and information program design and production for GTE, which includes over 400 downlinks. The network is used for product marketing, training, and various information programs.

VisNet employs a full-time staff of 46 and supports an additional 50 contract professionals. In addition, VisNet has established relationships with other organizations for maximum benefit to its customers. Budgets are determined per project and per resources.

These are examples of large departments. Almost every major company in the country uses video to some extent. Several companies have one-person departments. For example, Sentry Insurance in Stevens Point, Wisconsin, has a one-man corporate media department. Yet, it produces 40 programs a year, ranging from claims training programs to product information tapes for sales support.

Job opportunities for video professionals exist in virtually every industry. The ITVA membership directory is an excellent source of information on organizations currently using television.

VIDEO IN SPECIAL JOB SETTINGS

Although job titles and functions of video professionals may be similar from company to company, job demands vary considerably from industry to industry. This is so because each industry provides an environment or job setting that is different from that of the other. Some video professionals find it difficult to work in certain environments—hospitals, for example. It is important to have an understanding of the kinds of demands made by different environments before accepting a job in a particular industry. The ITVA has special interest groups (SIGs) classified by industry that serve as a forum for video professionals in common job settings. Job seekers should talk to ITVA members from many industry SIGs to get a broad perspective.

In the following pages, David R. Eatwell discusses the special demands of working in *legal video,* or video used in the process of law. Eatwell was founder and former president of the Houston, Texas-based Media Support Group, which specializes in providing media services to the legal profession.

Legal Video—Video in the Process of Law

The litigation process is by definition an adversary process. It is designed to mete out justice when each side of the matter in contention has made his or her case known as clearly as possible to the trier of fact, be that a judge or jury. The power and clarity of videotape lends itself to the concise, accurate depiction of the matter at hand.

Slowly assimilated into the legal process in many states, video can now be found in many areas and applications. For example, product liability and personal injury torts lead the way and still employ the widest range of video products. Early stages of case development might include videotape of an accident scene and interviews of witnesses, victims, treating physicians, or rescue workers. Later in the discovery process, depositions of key witnesses may be videotaped. A video crew may be dispatched to document the lifestyle of a victim who has been left with permanent disability or pain. Once the plaintiff's case has taken form, a documentary may be produced for presentation to the defendants in hopes of settling the case without the time and expense of delaying motions and trial.

As trial approaches, expert witnesses may turn to recreation of the accident on video to depict the event for the jury. This accident recreation can take many forms. The accident may be reenacted at the scene, with commentary and close-ups delineating the physical processes evident in the cause of the accident. Or the videographer may use computer-generated animation to portray the events leading up to the injury. Such documentation would then be presented to the jury as part of the testimony of the pertinent technical expert witness.

In very large cases, videotape is often used to prepare a witness for trial. A nervous witness may be drilled in how to handle grilling by a hostile attorney. A key witness with distracting mannerisms or poor speaking style may be coached in making a clear, credible presentation to the jury. Seeing oneself on video can be shocking to the uninitiated. Cases have been won and lost due to the preparation a witness received prior to trial.

Personal injury trials often include what has come to be called the "day-in-the-life" tape. This tape is edited from raw footage shot by a crew who follows the plaintiff through an average day. Verbal description of what it is like to be paralyzed is a far cry from the accurate depiction on videotape of normal activities of daily living. Though some decry such videotape as gruesome or inflammatory, the frustration of quadriplegia or the pain of burns healing cannot be fully described with mere words.

Legal video is still in a dynamic growth stage. The future holds no limits.

LEGAL VIDEO: LIMITATIONS

An overabundance of aspiring videographers has developed into a large body of legal videographers with little experience and relatively unsophisticated equipment. State, local, and national certification programs have recently arisen to solve the problem of prejudicial or incompetently produced video evidence.

Some types of legal video, such as settlement documentaries, call for all of the artistic expertise of a video producer. These products are not required to meet the rules of civil procedure pertaining to evidence. They are, however, subject to the scrutiny of a strong code of ethics.

Any video evidence to be shown in court must pass a strict test laid out by federal and state rules. Currently, few technical and procedural standards are set. The limitations primarily pertain to content.

Federal rules, along with some state rules, stipulate that the video equipment must be in good working order and that the operator must be qualified to render an accurate record of the evidence. Videographers who choose to enter this arena take on a responsibility for ethics, accuracy, and technical reliability that far surpasses the rest of the video industry. Indeed, failure in ethics, accuracy, or technical ability can lead to the videographer being sued. In no other area does a professional videographer have this type of exposure.

LEGAL VIDEO: SPECIAL SKILLS

The legal videographer must have the production expertise necessary to insure that the equipment is set up and operated correctly. In the case of a malfunction, back-up measures must be taken, which requires that the videographer be a good trouble-shooter. Because the legal videographer has little or no control over the video location, he or she must possess flexibility and the ability to deal with any situation. In general, the low glitz factor and high professional risk of this line of work call for a special type of person.

Getting into the legal video business is more than printing business cards and buying a camcorder. It is not wise for a novice videographer to consider independent work in this area due to the legal liabilities. A few years of experience with a corporate or commercial production crew that does lots of location work are a prerequisite. Frontline work is the only way to get the necessary experience.

There is no special place to gain experience in dealing with attorneys. Be prepared to wear a suit to depositions and court for tape playback. Be forewarned that many lawyers are suspicious of video and its practitioners. On the other hand, attorneys are demanding, with specifications and standards of accuracy laid out in law books and court decisions over the last 200 years.

In response to increased technical requirements, many states are now moving for controlled certification of legal videographers. As the uses of video in litigation abound, the threshold level for certification will rise. That is good. The average juror, who is an average citizen, watches about four hours of television daily. That level of viewer sophistication demands that legal video be better than home movies. A videographer who delivers less than professional quality video has performed a disservice to the client, attorney, and—more significantly—to the person represented in the case.

More than in other applications of video, legal video tangibly affects people's lives. The product may contribute significantly to the outcome of the case. Therefore, the decision to pursue a career in legal video is one that should not be made lightly.

HOW THE GOVERNMENT USES TELEVISION

The largest government-operated broadcast enterprise is Armed Forces Radio and Television Services (AFRTS). It operates more than 800 radio and television stations in 50 nations worldwide. In some cities, AFRTS operates out of conventional broadcast stations; in others it has closed-circuit television (CCTV) operations. AFRTS employs both military and civilian personnel to operate its television facilities and create programming.

Other government agencies at the local, state, and federal level, also use television extensively. Though the majority of them do not broadcast programs, most of them operate and maintain audio-visual equipment. Some departments generate their own programming, others contract with independent production companies to develop programs, and still others rent or purchase prerecorded videocassettes.

In 1991 the federal government spent an estimated $661.7 million on media of all types, according to Hope Reports, Inc. Approximately $200 million of the money was spent on television and video. The amount includes expenditures for salaries, off-the-shelf purchases, contracts for outside production, and capital equipment.

The Department of Defense (DOD) produces programs to train its military and civilian employees in specific tasks as well as to disseminate new policy information and medical or health-related data. The DOD spent $26.5 million on audiovisual activities in 1991.

At the state and local government levels, agencies utilizing video include public health, postal services, transportation, agriculture, and law enforcement. Some federal and state agencies use teleconferences and videodiscs. For example, the Center for Libraries and Education Improvement at the United States Department of Education commissioned the production of a six-sided videodisc project, entitled "The World of Work," to address the impact of new technologies on the job market of the 1980s.

Although the job market growth rate in the government sector is slow, employment opportunities in television and video at local, state, and federal agencies do exist. The Directory of U.S. Government Audiovisual Personnel, published by the National Audiovisual Center, is a good lead to federal agencies with television or video jobs. The directory can be obtained, for a fee, from National Audiovisual Center, 8700 Edgewood Drive, Capitol Heights, MD 20743-3701.

PRODUCTION AND FACILITIES COMPANIES

The explosion of the home video market into a multibillion-dollar industry, together with the widespread trend among the networks, independent stations, and cable companies to buy programs rather than produce original programming in-house, has led to the rapid growth in the number of production companies and postproduction facilities.

Large production companies such as Lorimar and MCA, produce half-hour programs for syndication or cable television. These programs include original situation comedies, game shows, made-for-TV movies and specials. Programs made for the home video market vary in style. Here are some categories with examples of titles and the program producer/distributor:

- Comedy: "Father Guido Sarducci Goes to College" (Vestron)
- Documentary: "The Cousteau Odyssey: The Nile" (Warner)
- War: "The World at War" series (26 volumes) (Thorn/EMI)
- Exercise: "Jane Fonda Workout" (Karl/Lorimar)
- How-To: "Wok Before You Run" (Embassy)
- Business & Industry: "Presentation Excellence with Walter Cronkite" (CBS/Fox)
- Children's Video: "Life With Mickey" (Walt Disney)
- Music Video: "We Are The World" (RCA/Columbia)
- Sports: "Chicago Bears Shuffle" (NFL)

While larger firms diversify the types of programs they produce in order to cater to a large audience, small firms tend to specialize. New York and Los Angeles are the headquarters for most of the companies that specialize in the production of commercials for advertising agencies. New York, Hollywood, Chicago, Dallas, Nashville, and Miami have a large number of companies specializing in producing entertainment programs. There are hundreds of small and large independent production companies that offer job opportunities to video professionals in every major city.

Most companies maintain a small full-time staff but flesh out the crew with free-lance help on a project-to-project basis. The *New York Guide,* which lists production companies, post production facilities, film and video labs, and sound/stage studios, is a good reference book for locating companies that hire video professionals.

Few independent production companies own the facilities necessary for the production and duplication of videotapes made for the consumer, corporate, and broadcast markets. Most rent the studios, remote trucks, and editing facilities of companies, which are often called production houses or postproduction facilities. Large companies, such as the Post House in Hollywood, California, and Unitel in New York offer the complete range of production and post production equipment and services. They are known as full-service companies. Some companies like Audio & Video International in New Jersey, specialize in services such as standards conversion. A good number of firms specialize in video duplication and distribution for the home entertainment and corporate/institutional markets. Some of these firms hire video specialists such as editors, computer graphic artists, color correctors, and music/sound recording artists. For a job at a production company or facility, you must be technically competent. The job emphasis in these firms is on engineering, technically skilled personnel, and marketing/account executives. The *Video Register* lists production/ post-production companies by state.

CHAPTER 4

PROGRAMMING AND PRODUCTION

Television is indeed one of the more dynamic industries in which to work. It offers a high-tech environment and the excitement of contact with interesting people. Work in the television business poses creative as well as technical challenges. But along with the stimulating work in the fast-paced television news programming area come back-breaking hours and erratic schedules. Nevertheless, the satisfaction of working with such a powerful news and entertainment medium makes for a rewarding career.

Since breaking into television is no easy task, it is essential to go about it in an organized and systematic manner. The first step is to understand how the industry is structured and identify the area of the business you want to be in. Most commercial and public television stations operate with five divisions: general administration, program-production, television news, engineering, and sales. Although there are a small number of on-camera or on-air positions, the majority of employment opportunities are in off-camera television jobs.

In the following pages, job titles, descriptions of responsibilities, and the qualifications required for the position are discussed briefly, so as to give you some idea of what television employees do.

TELEVISION NEWS

As the primary news source for most Americans, television stations strive to provide viewers with up-to-the-minute news, from early morning till late night. All the networks, and many local stations, interrupt programs with news breaks or update news by using crawl lettering at the bottom of the screen. The networks employ a relatively large staff and spend considerable sums of money for news gathering and delivery. Independent stations have smaller departments, but the emphasis on the news at any station can scarcely be overstated.

Job opportunities in television news and information departments are growing. This is due in large part to the increase in local television news coverage. With the growth of cable television, there are more programs aimed at audiences with specific or special information needs. While sports, weather forecasts, and traffic reports are regularly covered as specific news items, new special interest topics such as farm reports, consumer economics, health, and science are receiving local news coverage. Television news gathering and reporting is both an exciting and a demanding job.

Job Titles

News Director
Assistant News Director
News Producer-Director
News Desk Assistant
News Production Assistant
News Writer
Reporter
Anchorperson
Weather Reporter or Meteorologist
Sportscaster

Job Descriptions and Qualifications

A *TV News Director* has overall responsibility for a news team of reporters, writers, editors, and newscasters as well as the studio and mobile unit production crew. The job involves quick decision-making abilities, especially in situations yielding fast-breaking news. This is a senior administrative position, with responsibilities that include determining the events to be covered and how and when they will be presented in a news broadcast. This position calls for developing and administering the budget, monitoring and evaluating the performance of news staff, resolving production and technical problems, and coordinating news department activities with the programming and traffic or continuity departments.

A prerequisite for this managerial position is usually a college degree in broadcast journalism, mass communication, or political science, plus several years of experience in other television news positions or in radio news.

An *Assistant News Director,* also known as managing editor or assignment editor, is responsible for making news coverage assignments. The person in this position is responsible for supervising the news room and coordinating wire service reports, tape or film inserts, network feeds, and stories from individual news writers and reporters.

A *News Producer-Director* at some stations may be responsible for the same functions as the assistant news director, but the news producer-director has the additional responsibility of designating the technical crews for each assignment.

Preferred candidates for the positions of assistant news director or news producer-director usually have an undergraduate degree in journalism, mass communications, or political science plus several years of experience as a reporter or in other news functions. They also must have strong organizational skills.

A *News Desk Assistant* carries the responsibility of providing general assistance to the news department. This job entails general office duties such as answering telephones, opening and distributing the mail, delivering newspapers and magazines, filing scripts and correspondence, and distributing wire service copy to news writers and reporters. The person in this job may be called upon to transcribe portions of an interview and also collect routine information such as sports scores.

The basic requirement for this job is a high school diploma, preferably with additional college courses in business. This entry-level position is a good place to begin gaining the valuable experience that will help in qualifying for higher-level jobs.

A *News Production Assistant* has general duties such as organizing and maintaining the filing system for visual materials and locating library film or videotape footage for use in a newscast. The job may call for the preparation of all character generator information. In addition, the news production assistant may have to type out a lineup of the subjects scheduled to be covered in the program, including the timings of tape inserts.

This job generally requires a high school diploma. It is a good entry-level position for those aspiring to be news producers.

A *News Writer* is responsible for writing and editing news stories, continuity pieces, and introductions and descriptions that are used by anchorpersons in scheduled newscasts. A person in this position also writes narrative copy used as voice-overs for tape or film inserts.

News writers must have an undergraduate degree in journalism or mass communications, with a strong liberal arts background. One year's related experience or as a desk assistant is preferred.

A *Television Reporter* is a journalist who gathers news from various sources, analyzes and prepares the news, and reports it on the air. At a network station, a reporter who is assigned overseas is called a *Correspondent.* Reporters gather and investigate news through library research, telephone inquiries, interviews with key people, and news conferences and press briefings. They then develop the information into an understandable report, which is delivered on the air in a clear and concise manner.

Candidates for this position must be college graduates in journalism or political science, with a strong liberal arts background, and at least two years experience as a news writer. Preference is given to those with excellent writing and speaking abilities.

An *Anchorperson* or a newscaster hosts regularly scheduled newscasts. The person is responsible for reporting some of the major news items and for providing lead-ins for other stories. The anchorperson is the most visible element of a newscast. A person in this position must have a thorough understanding of news developments and the ability to analyze and interpret them for the viewing audience. Several nationally known anchors frequently research and write their own special news reports.

Candidates for this high-level position must have a distinctive personality and an ability to communicate with integrity and authoritativeness. A college degree in journalism or political science is required, along with many years of experience in other TV newsroom positions.

A *Weather Reporter,* also called weathercaster or meteorologist, reports the weather conditions and forecasts that are a part of regularly scheduled newscasts. The person in this position is responsible for gathering information from national satellite weather services, wire services, and other local and regional weather bureaus. The weather reporter works with the production crew and director to set up sophisticated visual equipment that will illustrate weather conditions.

This position has very limited employment opportunities. Major market stations and networks prefer college graduates in meteorology with good public speaking abilities.

A *Sportscaster* is responsible for reporting athletic and sporting events that are a part of regularly scheduled newscasts. A person in this position usually selects, writes, prepares, and delivers the sports news for each newscast. This may include interviews with sports personalities, live coverage of games broadcast by the station, and selection of video material for use in the sportscast. A college degree in journalism and a good knowledge of all sports are required for this position. Some newsroom experience also is valuable.

At a few major-market stations where there are several sportscasters, each sportscaster specializes in a particular sport. At these stations, there is usually a sports director who supervises and coordinates the team of sportscasters. Sports directors originate and develop special sports features and maintain contact with regional and national sports figures such as players, coaches, and managers.

Unions/Associations

News directors are usually members of the Radio Television News Directors Association (RTNDA). Membership in the association is also open to assistant news directors and other professionals in the field. Many news anchors and television reporters, several weather reporters, and some sportscasters have RTNDA membership. Assistant news directors with duties that include news writing or news reporting on the air may be members of the Writers Guild of America (WGA) or the American Federation of Television and Radio Artists (AFTRA).

At network stations and a few major market stations, WGA or AFTRA may represent news writers, news anchors, and television reporters for bargaining purposes. Some sportscasters at network stations and a few major-market stations belong to WGA or AFTRA. Sportscasters usually belong to the American Sportscasters Association.

Most desk assistants are not represented by a union, but at some major-market and network stations they are represented by the National Association of Broadcast Employees and Technicians (NABET).

Television reporters who cover Congress can belong to the Radio and Television Correspondents Association (RTCA).

PRODUCTION

Although it may be your dream to become a producer-director, you must remember that every television pro started at the bottom of the career ladder. So the best thing for you to do at the start is "everything." Get broad-based training and experience in as many production-related jobs as you can. At major-market television stations, job titles and functions are more streamlined and specialized. The smaller the station, the more different the kinds of jobs you will get to do.

Job Titles

Production Manager
Producer
Executive Producer
Unit Manager
Floor Manager
Production Assistant
Lighting Director
Graphic Artist
Associate Producer
Director
Assistant Director
Cartoonist
Cinematographer or Videographer
Videotape Editor
Scriptwriter

A *Production Manager* has overall responsibility for conceptualizing, designing, and developing programs for television. This function calls for coordinating and directing creative teams which include producers, directors, lighting directors, set builders, graphic artists, announcers, and on-air talent, to name only a few. In addition, the production manager has to supervise all local studio and remote production staff such as camera operators, production assistants, and floor managers. A person in this position has to prepare budget estimates as well as determine space and equipment needs.

A college degree in radio-TV, communications, or theater is a prerequisite for this management position. Some stations prefer candidates with graduate degrees and a minimum of five years experience as a television producer.

A *Television Producer* is responsible for planning and developing individual live or taped productions. The producer selects and directs the script, talent, sets, props, lighting, and other production elements. A person in this position is responsible for keeping productions within budgets and on schedule.

Producers are expected to have a liberal arts college degree with some training in TV, drama, or filmmaking, and at least three years experience as a director or associate producer.

An *Executive Producer* conceives, develops, and produces an entire series or some special production. Networks and large stations employ several TV producers who report to an executive producer.

A minimum of three years experience as a producer is essential for promotion to the job of executive producer.

An *Associate Producer* or assistant producer is the right-hand person of the television producer. A person in this position provides administrative and professional support in all aspects of production, from conceptualization to the final production. Associate producers help organize and implement production schedules and work closely with the operations manager in scheduling facilities and equipment.

Most employers hire associate producers with at least two years of varied television production experience plus a college degree in radio-TV or mass communications.

A *Director* or producer-director is responsible for rehearsing and directing a television program. This involves hiring the cast, plotting the camera shots, and determining production elements. These elements include equipment and engineering requirements as well as creative elements such as music, lighting, and sets.

Many TV directors have an educational background in theater and film. To be hired by a station in this position, an applicant must show at least two years production experience.

An *Assistant Director* is responsible for ensuring that all production elements—performers, equipment, sets, and staff—are ready for rehearsals and taping as scheduled. This person is also responsible for ensuring that slides, film, and tape inserts have been assembled and timed in the preproduction stage. On-location assignments call for arranging transportation, lodging, and facilities.

Two years of experience as a unit manager or floor manager are required for promotion to an assistant director.

A *Unit Manager* is primarily responsible for logistics and budget expenditures. With regard to logistics, a person in this position works on the setup, maintenance, and operation of all facilities and equipment during the preproduction stage. For remote productions, the unit manager arranges the rental of location, the delivery and setup of all equipment necessary for production, and the dismantling and return of all equipment upon completion of shooting.

Although this position does not call for college education, a minimum of two years experience in film or TV production is essential.

A *Floor Manager* coordinates all crew and talent activities in the studio as well as on location, in accordance with the director's instructions. It is the duty of the floor manager to follow along with the script and cue performers. Often, during production, the floor manager has to operate the TelePrompter, place props, and position easels. During production, the floor manager communicates the instructions of the director to the crew and talent, using hand signals.

A high school diploma and some college training in TV are usually required to become a floor manager. Those who wish to get this job in a major-market commercial station should have a college degree.

A *Production Assistant* does research, writes copy, schedules guests, interviews potential talent, and works on sets, costumes, and makeup—doing any and every job necessary at the time—to ensure a smooth flow of the production. This position is a stepping-stone for a career as a television producer.

Although the minimum requirement for this position is a high school diploma, the competitive nature of the business is making employers partial to candidates with a college degree.

A *Lighting Director* is responsible for achieving the lighting effects desired by the director in a production. Lighting directors use a variety of flood lights, spot lights, filters, and other accessories to provide the balance necessary for lighting a set. Some stations consider lighting to be a function of the engineering staff. In some cases, the duties of a lighting director are considered a production job.

Stations that consider lighting an engineering function require candidates for the job to have a high school diploma plus a second-class FCC license. Most lighting directors have at least two years of experience in related functions.

An *Art Director,* a *Graphic Artist,* and a *TV Cartoonist* are all considered part of the creative support team. This team designs and creates all art and visual materials used to enhance a television program. Duties include the production of visual elements such as photography, graphics, and animation art. The jobs may also call for the modification of existing sets and for scenic design.

Art directors usually have undergraduate degrees in commercial art and at least five years of experience in artwork for film or television.

A *Cinematographer or Videographers* also known as a camera operator or a cameraperson, is responsible for shooting all film and tape to be used in a program. Today, most television stations use portable electronic news gathering (ENG) cameras for location work, rather than film. A videographer must be adept at using many different types of cameras and recorders and must have creative technique.

An undergraduate degree in film and a minimum of one year of experience in the field is required for this position.

A *Videotape Editor* assembles and edits various pretaped segments, together with special effects and sound, into a finished program. Many editors specialize in the use of electronic paint and animation systems and digital video effects (DVE) devices. Videotape editing is a high-technology art form today. Opportunities for videotape editors also exist in production and postproduction facilities, some cable stations, and corporate studios.

The position of videotape editor usually calls for a person with experience in sophisticated editing techniques using computerized editing equipment.

A *Scriptwriter* develops a script as a blueprint for a production. TV and video scripts are typed into two columns, with audio on the right and video on the left. The audio column usually contains narration and dialogue as well as instructions for music and sound effects. The video column contains descriptions of sets, instructions to performers, and directions for camera shots and movements. Scriptwriters use a variety of styles such as documentary, talk-show, sitcoms, and drama. Often scriptwriters are called upon to write the audio portion after the documentary footage is shot.

Candidates with undergraduate degrees, usually in English literature, and some writing experience for film or television qualify for the job. They should also possess creativity and imagination.

Unions/Associations

In major markets and at some smaller-market stations, broadcasters are covered by collective bargaining agreements with one of three unions: The National Association of Broadcast Employees and Technicians (NABET); The International Brotherhood of Electrical Workers (IBEW); and the International Alliance of Theatrical Stage Employees (IATSE).

Videotape and film editors are represented by the IATSE locals, also known as the Motion Picture and Videotape Editors' Guild. Some film editors belong to the American Film Institute (AFI) and the Association of Independent Video and Filmmakers (AIVF).

Some art directors and graphic artists are represented by NABET. Many commercial television art directors and graphic artists are members of the Broadcast Designers Association (BDA). A few art directors are represented by United Scenic Artists.

While the majority of directors are not represented by a union, a few who work for networks and some independent production companies are members of the Directors Guild of America (DGA). Some assistant directors and unit managers working on TV network films are also represented by DGA.

There are no associations specifically representing the interests of producers or assistant producers. Many executive producers are members of the National Association of Television Program Executives (NATPE) or the National Academy of Television Arts and Sciences.

CHAPTER 5

ENGINEERING AND TECHNICAL SERVICES

Good engineering and technical people are in great demand. Spurred by the rapid growth of the consumer electronics and computer industries, there is a real shortage of qualified technical personnel in the television industry.

The increasing number of video facilities in nonbroadcast industries and the steadily evolving cable industry account for the growth in opportunities for those with technical school training and television engineering experience. The quality of video and audio signals transmitted by a station depends on the engineering and technical staff. Also important are the performance and reliability of professional video equipment used in production or postproduction facilities and corporate video departments. As the level of sophistication of equipment constantly increases, so do the demands that technicians and engineers upgrade their skills and become proficient in handling new technology.

Engineering and technical job titles and functions differ not only on the basis of station size but also as the employee moves from a major market to a smaller market. Vertical job mobility is a common way to advance in a technical career. For example, an engineering technician with a year or two of experience at a station may get promoted to the position of master control engineer. An engineering technician also can move further up the career ladder by joining a smaller station or moving to a smaller market.

Lateral moves are more common among engineering management personnel. Some chief engineers advance their careers by moving to a smaller market in a job with the same title but with more responsibility and at a higher salary.

Job titles and functions at commercial and public TV stations tend to be similar, but the nature of the functions at a cable TV station are often different.

Working in a technical or engineering job at a facility in a nonbroadcast industry is similar to working at a small commercial station in a very small market. Operations and maintenance functions are combined, and often the equipment is less sophisticated.

Here are some of the job titles for engineers and technicians in the television/video industry:

JOB TITLES

In Commercial and Public TV

Chief Engineer
Assistant Chief Engineer
Engineering Supervisor
Maintenance Engineer
Transmitter Engineer
Audio-Video Engineer
Videotape Engineer
Master Control Engineer
Engineering Technician
Technical Director
Camera Operator

In Cable TV

Chief Technician or Engineer
Bench Technician
Maintenance Technician
Service Technician
Trunk Technician
Installer

In Nonbroadcast Industries

Chief Engineer
Maintenance Engineer
Maintenance Technician
Media Technician
Production Technician
Video/Audio Engineer
Video/Audio Technician

JOB DESCRIPTIONS AND QUALIFICATIONS

Commercial and Public TV

A *Chief Engineer,* sometimes known as the director of engineering, is ultimately responsible for all of a television station's technical facilities and services. This person is also responsible for the supervision of all engineering and operations staff. The position calls for long-range facilities planning, budgeting, and purchase of equipment. The chief engineer is charged with the responsibility of ensuring that all station operations are in compliance with Federal Communications Commission (FCC) regulations and other applicable local, state, and federal laws. Therefore, a thorough knowledge of FCC regulations is a prerequisite.

A chief engineer must have an FCC first-class license. The preferred educational background for this position includes an undergraduate degree in electrical engineering and technical training in broadcast engineering. Usually, five years of experience as an assistant chief engineer is one of the prerequisites for the position.

An *Assistant Chief Engineer* is responsible for overseeing the day-to-day technical operations of the station. This includes scheduling facilities as well as the technical staff. The job calls for overall staff training in the operation and maintenance of new equipment. The person in this position is expected to supervise equipment modification and replacement, conduct transmission tests, and oversee inventory of parts and other supplies.

Qualifications for this high-level job include a high school degree (though an undergraduate degree in electrical engineering is preferred), plus technical training, a first-class FCC license, and at least five years of television engineering experience.

An *Engineering Supervisor* is directly responsible for the proper operation and maintenance of equipment on a day-to-day basis. A key duty is the supervision of the line staff, from videotape engineer-editors to audio-video engineers. This position calls for maintaining and reviewing technical reports as well as diagnosing and correcting technical problems.

A minimum of a high school education plus one or two years of technical school training is expected of an engineering supervisor. Candidates with a first-class FCC license plus at least two years of experience as an audio-video engineer are preferred.

A *Maintenance Engineer* is responsible for the maintenance and repair of all television equipment at a station or production facility. The job calls for ensuring that equipment meets the design specifications. Therefore, a person in this position must be adept at using various test equipment to gauge the performance of production and transmission hardware. A maintenance engineer must be able to handle complex repairs and equipment modification.

A high school diploma, a second-class FCC license, and at least one year as an engineer technician will qualify candidates for the job. Most maintenance engineers at TV stations take the examination for the first-class FCC license.

A *Transmitter Engineer* operates and maintains a television transmitter. Transmitter engineers are often known as field service engineers, since the transmission tower and antenna are in some instances located in a remote spot away from the main studio. Because this job calls for the continual monitoring of all incoming satellite, network, or regional broadcasts, a transmitter engineer is assigned to various shifts during a broadcast day.

Typically, transmitter engineers have a high school diploma, training in broadcast engineering from a trade school, skills in the maintenance of electronic equipment, and a first-class FCC license.

An *Audio-Video Engineer* has the overall responsibility of operating all electronic audio and video equipment. Audio-video engineers are an essential part of the production team because they are charged with the responsibility of achieving the best possible sound and image quality. An audio-video engineer works in the control room during production, in a remote truck during an on-location shoot, and in the editing suite during post-production. Audio-visual engineers, audio technicians, and video technicians are employed at commercial and public TV stations, production and post-production facilities, and television or media departments in nonbroadcast industries.

Although most employers require audio-video engineers to have a second-class FCC license, many TV stations prefer candidates with a first-class license. The position usually calls for a high school diploma, some technical training, and experience as an engineering technician.

A *Videotape Engineer-Editor* is responsible for the setup and proper operation of all videotape machines. The person in this position cleans, makes adjustments and alignments, and sets up videotape recorders, editors, and players prior to operation. With the relative ease of use and maintenance of today's sophisticated recording and editing videotape systems, the videotape engineer-editor is not actually required to be an engineer. At a production or post-production facility, people hired for this position are responsible for *dubbing* (duplicating) tapes and assembling and editing commercials and other promotional spots. Since the emphasis today is more on videotape editing than on engineering, people from production backgrounds are moving into this position.

A minimum of a high school diploma and some technical training are prerequisites for the job, along with experience in computerized videotape editing or a year as an engineering technician. Some stations require videotape engineer-editors to have a second-class FCC license.

A *Master Control Engineer* ensures that all of a television station's scheduled program elements, such as on-location feeds and prerecorded segments, are smoothly transmitted. This job calls for coordination of the output of several departments as well as the handling of electronic machines. The person in this position operates the master control switcher and audio console during station breaks.

A master control engineer is directly responsible for ensuring that all transmissions meet FCC requirements, thus candidates with a first-class FCC license, experience as an engineering technician, or some technical training upon completion of high school are preferred.

An *Engineering Technician* is responsible for a variety of engineering duties, such as the maintenance and setup of cameras, video and audio tape recorders, signal routing switchers, and testing devices. A person in this position at a TV station usually reports to an engineering supervisor. At facilities in nonbroadcast industries, an engineering technician or operator technician reports to a technical director or chief engineer, depending on the size of the facility. After gaining experience for a year, engineering technicians usually get promoted and trained in an area in which they have shown talent and interest.

A prerequisite for the job is a high school diploma and some technical training. Candidates with a second- or third-class FCC license are preferred.

A *Technical Director* works closely with the program's director in determining the specific requirements of a production. The person in this position is responsible for operating the production switcher. He or she also directs studio and control room technical staff and—according to the director's instructions—guides camera operators.

For this job, applicants must have a high school diploma, some technical training and experience as a lighting director or camera operator, and a thorough understanding of both the production and technical aspects of television programming. Many stations require that candidates for the position have a second-class FCC license.

A *Camera Operator* is responsible for setting up and operating television studio and Electronic News Gathering (ENG) cameras. While camera operators are considered part of the engineering department at some stations, they are part of a production or news department at other stations. Many camera operators are independent cinematographers and videographers.

TV stations hiring camera operators require that candidates have a high school diploma and training in photography, film, or video, plus some experience in television production.

Cable TV

A *Chief Technician* in a cable TV company is responsible for the supervision of system installation, transmission, and operation. At a multichannel multipoint distribution service (MMDS) or at subscription television (STV) companies, a chief technician is sometimes called a chief engineer. This job calls for both technical and administrative skills. Engineering responsibilities include supervision of the construction and installation of antenna towers, dishes, amplifiers, and other signal-processing hardware. Administrative functions include evaluation and purchase of all equipment, procurement of permits from telephone and local electric companies, and supervision of all technical staff which can vary from three to over 20 employees. Chief technicians or chief engineers are expected to have good design skills.

This position usually calls for at least five years of broadcast or cable engineering experience, a first-class FCC license, and preferably college course work in electrical engineering.

A *Technician* in cable TV usually carries a more specific job title that reflects a specialized position, such as bench technician, maintenance technician, service technician, or trunk technician. Each technician is responsible for the installation, maintenance, and repair of a certain type of equipment. For example, bench technicians work on subscriber converter boxes, while maintenance technicians work on damaged cable between telephone poles, testing equipment, transmission antennae, and signal scrambling units. Service technicians repair receiving antennae and subscriber signal descrambling units. Trunk technicians work in the field, repairing main cable lines.

Most service technicians are expected to have a second-class FCC license. Several cable companies also expect trunk technicians to have a second-class FCC license. A high school diploma, and some trade school training in electronics are general requirements for these positions.

An *Installer's* job function varies according to the type of company the installer works for. The position at a cable company calls for wiring the receiving end of the television signal from a telephone pole to the connector box to the TV set. At an STV or

MMDS company, an installer is responsible for attaching an antenna to the roof of the subscriber's house. The installer connects the antenna to the descrambling unit and to the subscriber's television set. If an installer works for an independent contractor, the job may call for digging trenches, relocating telephone poles, and laying down cable between poles.

A high school diploma and demonstrable mechanical aptitude and technical ability are required for this entry-level position.

UNIONS AND ASSOCIATIONS

In commercial TV, many technical directors, engineering technicians, videotape editor-operators, audio-video engineers, transmitter engineers, and maintenance engineers are usually represented by the National Association of Broadcast Employees and Technicians AFL-CIO (NABET) or the International Brotherhood of Electrical Workers (IBEW). Individuals with the same job titles in public TV usually are not represented by a union. However, at some major-market community stations, some individuals with these titles are members of one of these unions.

The titles of engineering supervisor, assistant chief engineer, and chief engineer are considered management positions and therefore are not represented by a union. Individuals who may have been members of a collective bargaining agent or union such as NABET or IBEW let their membership lapse or become inactive members upon assuming a management position.

Individuals with engineering or technical services management positions usually belong to a professional association to increase their technical knowledge, share mutual concerns, and advance their careers. Some broadcast engineering executives belong to the Society of Broadcast Engineers (SBE), or the Society of Broadcast and Communications Engineers (SBCE), or the Society of Motion Picture and Television Engineers

(SMPTE). Some engineering executives have membership in more than one association.

There are no unions that represent installers, bench technicians, maintenance technicians, service technicians, or trunk technicians working in the cable television industry. Chief technicians or chief engineers are usually members of one or more of the following organizations: the Society of Cable Television Engineers (SCTE), Community Antenna Television Association (CATA), Association for Maximum Telecasting (AMT), SMPTE, or SBE.

Video service technicians, maintenance engineers or technicians, and media technicians who work in nonbroadcast industries are not represented by unions. Maintenance engineers are usually members of the Institute of Electrical and Electronics Engineers (IEEE). Video-media service and maintenance technicians usually belong to one of the following: the National Association of Television and Electronic Servicers of America (NATESA), the National Electronic Service Dealers Association (NESDA), the International Society of Certified Electronic Technicians (ISCET), or the Association of Audiovisual Technicians (AAVT).

CHAPTER 6

SALES AND MARKETING

It has been said, "If all you want for dinner is a pizza, go do production; if you'd like spaghetti and meat sauce, then get into service; but if you'd like to afford a gourmet meal, sales is your best bet." Well, it took a salesperson to make a statement like that! It is generally true, however, that the sales and marketing area is the most lucrative in this industry.

Television stations derive revenue from the sale of on-air time. The three main sales sources are national and regional advertisers, local businesses, and network sales (in the case of network affiliates airing network commercials and programs). The income received by each station varies with the size of the station and the market in which it operates. By and large, national advertisers spend far more money in major markets than in smaller ones.

BROADCAST TELEVISION

When selling time, a television salesperson is in reality selling the demographics of the audience that watches the station's programs. Media buyers are interested in knowing how many men, women, and children can be reached with a TV commercial at a particular time. Therefore, television salespeople rely heavily on research and audience demographics. The price of advertising time is proportional to the number of people who can be expected to watch it. A TV salesperson must have a thorough understanding of the station's programs and sales policies. A salesperson must also be able to make complex cost computations and propose a commercial schedule to a potential advertiser.

A television station's local sales staff makes calls on advertising agencies and, if need be, on an agency's client. Many large TV stations in major markets engage the services of a national sales representative firm to function as the station's out-of-town sales force. These firms work on a commission basis. TV sales employment opportunities are available at TV stations as well as national sales representative firms.

CABLE TELEVISION

Cable companies get most of their income from subscriber revenues. Hence, many cable marketing jobs involve door-to-door or telephone sales, especially where new systems have been installed. Many of these jobs pay a commission only on sales made. At some companies, the position of marketing or sales manager is an on-staff position. The job calls for sales promotion activities, processing customer requests, and getting CATV service installed.

The largest job sector in the area of cable TV is in customer service. Companies hire customer service representatives at the local level. A person in this position is responsible for resolving current customer complaints and upgrading current customers to better service plans. Job titles and functions in cable TV vary considerably from company to company.

NONBROADCAST ORGANIZATIONS

Manufacturers and dealers of television equipment hire sales and marketing people to promote and sell their products. In this environment, sales and marketing personnel need to have a thorough understanding of the company's products, its applications, and the special needs of its user groups. They must also have the ability to strengthen marketing strategies so as to give the company's product maximum visibility.

JOB TITLES

Broadcast Television

General Sales Manager, Sales Director, or Advertising Sales Manager

National Sales Manager or Assistant Sales Manager

Advertising Salesperson, Time Salesperson, or Broadcast Salesperson

Sales Coordinator, Traffic Coordinator, or Sales Assistant

Cable Television

Marketing Director

Sales Manager

Salesperson

Nonbroadcast Organizations

National Sales Manager

Regional Sales Manager

District Sales Manager

Sales Rep, Salesperson, or Sales Agent

JOB DESCRIPTIONS

Broadcast Television

A *General Sales Manager* has the overall responsibility for generating all advertising revenue. In conjunction with the general manager of the station, the general sales manager establishes the station's advertising policies. The job calls for establishing advertising rates, preparing a sales forecast, developing a sales strategy, and supervising a sales force.

An undergraduate degree in marketing, advertising, or business administration is preferable. Candidates with a minimum of five years of experience in broadcast advertising sales have a competitive edge.

A *National Sales Manager* has responsibilities including supervision and servicing of all national accounts; the hiring, training, and supervision of sales staff; establishing and monitoring quotas for each advertising salesperson; and the preparation and maintenance of sales records.

An undergraduate degree in marketing or advertising is preferred; a minimum of a three-year successful sales track record in television advertising sales is required for a promotion to this position.

An *Advertising Salesperson* is involved in the day-to-day selling of advertising air-time. Each salesperson is assigned a specific territory, such as local sales or national sales, or to a task such as obtaining sponsors for particular programs. Duties include calling on advertising agencies and businesses. Advertising time is sold in 10- or 30-second spots. In selling a time spot, the advertising salesperson is expected to match the station's available advertising air time with a client's advertising need. To be effective, a salesperson must understand specific audience shares and rating.

Candidates with an undergraduate degree in marketing or advertising and with a minimum of one year of retail or print advertising experience are preferred.

A *Sales Coordinator* serves as an administrative assistant to the sales manager in coordinating all advertising activities. Duties include maintaining the master sales files and schedules, writing advertising orders, and maintaining a schedule of available air time. A person in this position may also be responsible for coordinating sales activity with the traffic continuity department and the production department.

A minimum of a high school diploma, with some general office administration or secretarial training, is preferred. Because most of these positions require the use of a computer, computer literacy is required for this position.

Cable Television

A *Marketing Director* is responsible for the overall marketing, promotion, publicity, and advertising activities necessary to increase the number of cable subscribers. A person in this position is responsible for the supervision and training of sales staff. He or she establishes rates, discounts, and special offers for program services.

Most cable TV marketing directors have an undergraduate degree in marketing or advertising and at least three years of experience in telecommunications sales and marketing.

A *Sales Manager* has direct responsibility for all sales. Duties include assigning salespeople to particular accounts and supervising their activities on a daily basis. A person in this position has to keep all sales staff informed of changes in service, prices, discounts, special offers, and organizational policies. The main task of a sales manager is to work closely with the sales force to aggressively sell the station's services.

A *Salesperson* for a cable television station is responsible for signing up customers for a station's program services. Most sales are made on the telephone and through door-to-door selling.

A high school diploma and some retail or telephone sales experience are necessary for this job.

Nonbroadcast Organizations

Sales and marketing jobs in nonbroadcast organizations are available at consumer video stores, professional video dealerships, the distribution and sales divisions of television hardware manufacturing companies, and at production and post-production facilities. At production and post-production facilities, an account executive sells studio or facilities time.

While many of the sales aspects of the jobs listed in this category are common, job titles and specific duties may differ according to product and size of the company. At professional video dealerships and at manufacturing companies, the career path for salespeople starts at the sales representative level. The career ladder runs from sales rep to district sales manager, to regional sales manager, to national sales manager, and onward to management positions such as marketing and sales manager and general sales manager. Most positions in hardware sales at manufacturing companies call for prior experience either in video retailing or at dealerships.

A *National Sales Manager* is responsible for the sales of a certain group of products nationwide. Duties include writing proposals and marketing bulletins regarding the competitive edge of their product; determining special pricing; hiring, training, and supervising of regional sales staff; developing marketing strategy; and making media buying decisions regarding print advertising.

A *Regional Sales Manager* is responsible for supervising the activities of district sales managers. Their duties include achieving quotas for the region; setting up exhibits at regional trade shows and at local video association meetings to introduce new equipment to dealers and end-users; and administration of dealer contracts and regional office staff and facilities.

A *District Sales Manager* usually has several dealerships as accounts within a specified territory. A district sales manager's duties include aiding dealers to sell more products by identifying the unique applications of a product that meet client's production requirements. The job may entail hauling heavy equipment or crawling behind racks of equipment to set up and adjust products for demonstration. District sales managers follow up on sales leads. They are responsible for achieving their quotas on a monthly, quarterly, half-yearly, and annual basis.

SELLING SKILLS AND JOB DEMANDS

Contrary to what most people think, you do not have to be "born" a salesperson to be in sales. There are many sales skills that can be acquired by those who do not have them innately. However, there are several personal attributes which are prerequisites for a successful sales career.

Experts on selling suggest that the primary attribute of successful salespeople is a strong desire to make money and a tremendous confidence that they will succeed. In addition, they must be self-motivated and have a positive self-image. Sales jobs require courage and perseverance, a pleasing personality, and honesty and integrity.

According to John Rhodes, a marketing manager for Sony Communications Products Company, "It is important to have a thorough knowledge of your product, its end-users, and competitive products in the market. Equally important is your ability to talk about it in an intelligent yet comprehensible manner."

To be in sales, you must be articulate and persuasive in speech. You must have the ability to listen attentively without interrupting the speaker. You must be able to write effectively. To be successful in sales, according to Rhodes, "You must have excellent organizational skills. The successful ones use a pre-call planning method to prepare for sales calls and are very thorough with follow-up correspondence. They are very disciplined with their paperwork—be it writing a sales order or filing an expense report."

According to Marc Feingold, director of sales and marketing at Nimbus Records, Inc., "The essence of sales is service. Give service—honest, good service. Get involved with your customers. You have to understand their business and find new ways to serve them. You have to be willing to go out of your way to help your customers, even if it means driving hundreds of miles out of your way. You don't have to be a good storyteller or be able to tell jokes. But you must have a sense of humor and the ability to laugh, as many difficult situations may call for it."

You have to be good at getting out of bad situations. If you have a problem, you must face it right away. Salespeople always have to maintain a positive disposition even in the face of adversity.

As attractive as a sales job may appear, it can be very demanding and stressful. Given a territory, your life is pretty much on the road. Living out of a suitcase is always difficult, and it can be physically tiring. With much of your time away from home, it can also be difficult on your family.

Much of the stress associated with selling is related to achieving sales targets. Since most of the sales jobs are compensated, in part, on a commission basis, the push to achieve quotas can be stressful, but the monetary rewards can be substantial.

CHAPTER 7

WORKING CONDITIONS AND EMPLOYEE COMPENSATION

Although television is "show biz," it does not necessarily mean that working in this field will bring big bucks. Working in a television studio, be it in a broadcast or nonbroadcast environment, does offer the excitement and some of the glamor of show business. It will also offer you a challenging career. But only a few career television professionals really get rich at it. Most make a reasonable living, and some do see difficult days. The majority, however, gain career fulfillment and job satisfaction that far outweigh their paychecks.

While studio working conditions are usually plush and pleasurable, location assignments may be tedious for those who are ill-prepared to handle inclement weather conditions and other production difficulties posed when on a remote shoot. Again, while some people thrive on tight deadlines and the fast pace of a newsroom environment, others may find it nerve-wracking and taxing.

It is necessary for you to have a good understanding of the working conditions and compensation—salary and fringe benefits—before embarking on a career in this field. Several industry surveys will provide you with the information you need.

SURVEYS

A number of organizations report the employment figures, salaries, and fringe benefits of TV and video employees, based on a survey of their members or on a certain segment of the industry. These surveys, usually conducted on an annual basis, provide a method by which those working in the industry can compare their income with national norms. The surveys will give you a good indication of the number of opportunities available on the basis of job titles as well as the salary range of employees currently holding positions with those job titles.

Because salaries reflect the general health of the industry to some extent, these survey results should provide you with some insight into how those working in the industry are faring. You'll be able to judge which segments of the industry are doing well or facing hard times. In addition, a number of survey reports will point out trends in salary increases and employee benefits.

Here is a list of annual survey reports you should check:

- *Broadcast and Cable Industry Trend Reports,* Federal Communications Commission.
- *Radio and Television Employee Compensation and Fringe Benefits Reports,* National Association of Broadcasters.
- *Annual Membership Salary Survey,* International Television Association (ITVA). It includes salaries within job descriptions, geographical regions, nonbroadcast industry classifications, and major departments. It also provides salary information based on length of experience in corporate television, on department operating and capital budgets, and on income information of freelance professionals. The survey results are available to ITVA members without charge and to nonmembers for $50.
- *Annual Salary Survey,* and *State of the Industry Survey,* published by *Broadcast Engineering* magazine. The report is based on responses to questionnaires mailed to recipients of

the magazine, selected on an "*n*th name" basis. It includes profiles of: 1) management staff; 2) engineering and technical staff; and 3) operations staff. Each profile includes salary level, salary increases during the past year, fringe benefits received, number of years in present job, years in broadcast industry, and education level.

- *Salary Survey,* published by *Video Systems* magazine. The report is based on responses to questionnaires mailed to readers on an "*n*th name" basis. Information presented includes profiles of staff in management, production, technical areas, and training. Data is classified on the basis of respondent's employment setting. Categories include teleproduction, TV station, corporate, medical, education, government, and cable TV.

The NAB Survey

The following three tables are from the 1991 "Television Employee Compensation and Fringe Benefits Report" compiled by the National Association of Broadcasters (NAB). Questionnaires requesting salary information were mailed to all U.S. commercial television stations. Of the 1,079 stations surveyed at that time, 554 stations completed the survey.

The first table gives salary information on department heads; the second gives salary figures for support staff; the third shows annual compensation for sales personnel.

1991 NAB/BCFM Television Employee Compensation and Fringe Benefits Report

Department Heads	*Average Base Salary*	*Average Bonus*	*Median Base Salary*	*Median Bonus*
General Manager	113,347	20,583	100,000	9,800
Asst. General Manager/ Station Manager	69,609	6,879	63,000	0

Department Heads	*Average Base Salary*	*Average Bonus*	*Median Base Salary*	*Median Bonus*
Operations Manager	45,090	1,515	40,000	0
Program Director	43,922	1,815	38,000	0
Director of Engineering/Chief Engineer	47,741	1,459	44,720	0
News Director	58,152	2,349	48,720	0
Marketing Director	53,468	2,992	45,112	0
Promotion/Publicity Director	35,389	1,099	31,340	0
Production Manager	34,016	903	31,801	0
Business Manager/ Controller	44,396	2,138	40,500	0
Traffic Manager/ Supervisor	27,365	417	25,334	0
Community/Public Affairs Director	32,848	481	30,000	0
Art Director	31,601	125	29,013	0
Research Director	38,702	1,065	34,425	0
Human Resources Director	37,684	1,084	33,085	0

Support Staff	*Average Number Fulltime Employees*	*Annual Compensation ($)*		*Average Starting Compensation ($)*
		Average	*Median*	
Operator Technician	9	22,136	19,500	16,017
Maintenance Technician	4	28,280	27,210	21,372
Technical Director	4	24,705	20,453	19,790
Floor Director	2	22,594	19,597	17,059
Film/Tape Editor	3	22,379	20,000	17,496
Film Director	1	24,310	23,400	18,336
News Anchor	4	64,763	41,000	33,725
News Producer	4	26,169	24,371	21,515
News Reporter	7	29,035	21,825	21,073
News Photographer	8	22,637	19,064	17,357

Support Staff	*Average Number Fulltime Employees*	*Annual Compensation ($) Average*	*Annual Compensation ($) Median*	*Average Starting Compensation ($)*
Sportscaster	2	46,066	31,900	27,380
Weathercaster	2	48,158	36,660	27,974
Assignment Editor	2	27,109	26,500	22,063
Production Assistant	3	17,734	16,000	14,699
Producer/Director	4	26,410	23,806	20,131
Staff Artist	2	23,539	22,000	18,960
Traffic/Computer Operator	3	17,457	16,484	14,539

Sales Personnel

	Account Executive	*General Sales Manager*	*Local Sales Manager*	*National Sales Manager*	*Co-op Coordinator*
Number	7				
Annual Compensation ($)					
Average	44,565	91,848	70,487	74,431	41,305
Median	40,000	85,000	67,000	73,000	32,000

SALARIES IN NONBROADCAST ORGANIZATIONS

Television professionals who work in nonbroadcast organizations will tell you that it is not a road to riches or fame. It certainly offers an exciting field of endeavor. But with it comes long hours of hard work and often more complaints than accolades. While employment at a broadcast television station may offer you some glamor, most TV professionals agree that there is very little, if any, job security. It is often said that in broadcast television, you are only as good as your last ratings. However, those working in corporate or organizational video seem to have a better sense of job security. Of course, no organization offers an absolute guarantee of job security.

The ITVA annual salary survey results in the following table are representative salaries of people working in nonbroadcast organizations.

ITVA 1991 SALARY SURVEY
SALARIES WITHIN JOB DESCRIPTION
TABLE #1 GENERAL

Position	*# Resp.*	*10% Tile*	*25% Tile*	*50% Tile*	*75% Tile*	*90% Tile*
Manager	257	29,000	35,000	43,000	54,500	72,000
Supervisor	123	25,892	30,000	37,000	45,000	50,000
One Person Operation	157	20,000	25,200	32,000	40,000	48,000
Producer	296	23,500	28,000	34,000	40,000	50,000
Assistant Producer	17	18,500	20,400	23,748	31,000	37,500
Director	39	20,000	25,000	30,000	40,404	70,000
Production Assistant	19	10,000	13,000	16,900	21,070	27,000
Writer	35	15,783	28,000	33,800	39,600	70,000
Editor	20	18,000	24,500	30,000	43,000	55,000
Audiovisual Specialist	89	19,700	22,500	26,250	31,500	37,000
Engineer	18	20,000	27,400	36,900	48,679	60,000
Technician	23	17,000	21,400	26,016	30,456	40,000
Video-grapher	25	8,050	16,030	25,000	30,000	37,000
Sales/ Marketing	56	22,500	28,000	38,500	50,000	65,000
Professor/ Instructor	33	22,000	29,000	33,000	39,000	45,000

Student (Full-time)	7			10,000		
Other	30	10,000	21,300	33,800	42,000	50,000
Overall	1,246	21,000	26,800	34,500	43,000	55,530

The survey report also includes salaries within job descriptions in separate geographic regions. Differences in salary for the same job description or title are noticeable. For example, in the 1991 survey report, managers who made $48,500 per year were in the 50th percentile group of the 30 managers who responded to the questionnaire from Region II (comprised of Delaware, New Jersey, eastern Pennsylvania, and eastern New York). In contrast, managers in the 50th percentile group in Region V (western Pennsylvania, western New York, Ohio and Michigan) earned only $42,000 per year. There were 27 managers who responded to the survey questionnaire from that region. The ITVA survey also reveals differences in salaries based on job description or title within organizational classification (aerospace, medical, or insurance, for example). A producer at a hospital media department, for instance, may not have a salary comparable to a producer at an aerospace company.

EMPLOYEE BENEFITS

Benefits given to employees vary considerably, depending on the segment of the industry (broadcast, cable, or nonbroadcast), the market size, and the size of station or organization. Since this is not an industry where everyone makes "big bucks," the benefits package accompanying salary compensation is of considerable importance to most full-time, on-staff employees. Experts in the field state that in some segments of the industry, you can negotiate a better fringe benefits package at the management level or in sales and marketing positions.

Employee benefits typically received by television/video employees include:

Medical insurance

Dental insurance

Life insurance

Sick leave

Vacation

Stock purchase plan

Profit sharing plan

Savings plan

Pension plan

Bonus

Trade show, convention, or seminar expenses paid

Tuition refund plan

Automobile, (furnished by the station or company)

There is no standard plan offering the benefits listed here. But the majority of television or video employees receive medical coverage and paid sick leave. Bonuses and pension plans are considered basic fringe benefits. Several organizations also offer profit sharing and stock purchase plans. Although an increasing number of organizations offer tuition refund, perks such as company-furnished automobiles are not common.

CHAPTER 8

EDUCATION: ACQUIRING SKILLS AND IN-DEPTH KNOWLEDGE

There is no ideal pathway to a career in television or video. A few lucky people start early, for example, when their uncle lets them borrow a camcorder to tape the community football game. Some get a head start in high school, usually lugging around and helping set up AV equipment for their favorite teacher. Many people go to college and major in journalism or broadcasting. Still others get a liberal arts degree and learn all that they know about television on the studio floor in their first job. Very few specialize in telecommunications at a graduate school and then embark on a career in television.

No matter what the path of learning may be, one can learn a great deal from those with long and wide-ranging experience in the field. The important thing is to start early and learn as much as possible from as many pros as possible.

HIGH SCHOOL PREPARATION

You can start preparing for a television career while you are still in high school. Students who develop good study habits are laying a foundation for the future. Discipline and research skills are essential for a number of television jobs. While most of the subjects you study in school will serve you well in the future, special attention to writing, spelling, English, history, geography, social studies, and mathematics will be of most practical value.

Channel some of your extracurricular activities toward developing communications skills, both written and spoken forms. Join a drama group, literary club, or debate team to improve your public speaking abilities. Your participation in a drama group may involve assisting with set construction or setting up lights.

Read as much as you can about radio and television. Books on broadcasting are available at most libraries. Visit your local radio and television station and arrange to talk to the people working in production. Ask them questions about their jobs, what they like and don't like about their work.

Some high schools operate a closed-circuit TV (CCTV) studio. For example, Beech Grove High School in Indiana has a three-camera studio with 3/4-inch editing capability. The school produces two programs a day, covering school and community activities. Beech Grove feeds the local cable company with programs of high school games and other sporting events.

In the Indianapolis area alone, three high schools offer basic and advanced television production courses to junior or senior students. If your high school has such a program, take it. If your school operates a CCTV studio, volunteer some time to assist the staff. Do anything they need you to do, and learn as much as possible.

DEGREE PROGRAMS

A college degree is not required for some jobs in broadcast television. However, because the industry is so highly competitive, candidates with a college degree find it easier to get a job. For most supervisory and management positions, higher education is required.

In response to the sixteenth survey conducted by the Broadcast Education Association (BEA), of American four-year colleges and universities which offer degrees or coursework in broadcasting, 265 schools offering an undergraduate degree reported a total of 34,646 junior and senior students majoring in broadcasting. Upon graduation, a good number of these people will be seeking employment in the television industry. So, as you can see, the competition is stiff.

While it is important for you to learn the skills and techniques of radio and television broadcasting, it is imperative that you receive a well-rounded education. Over and over again, television executives interviewed for this book stressed the need for students to obtain a comprehensive liberal arts background.

According to Erik Barnouw, Professor Emeritus, Columbia University, "An encouraging tendency in television curricula, and to which students have warmly responded, is a choice of double majors. The student who is a specialist in television but also in history—or also in government, anthropology, sociology, psychology, public health, or international relations—is assuming growing importance. Such broad-based, interdisciplinary approaches help fill in more of the television 'window' and should certainly be encouraged."[1]

Although the depth and quality of education at broadcasting and telecommunications departments varies considerably, some basic standards are observed by all schools accredited by the Accrediting Council on Education in Journalism and Mass Communications (ACEJMC). Before selecting a college or university, be sure to get as much information as possible from a number of the institutions you are interested in. You will need to get answers to some basic questions: What is the emphasis of the department's curriculum? Are the majority of courses offered theoretical or

---

[1]Erik Barnouw, "Foreword On Television," *The American Film Institute Guide to College Courses in Film and Television.* New York: American Film Institute, 1975. p. ix.

practical? Have the faculty members worked professionally in radio or television? Does the school have a well-equipped television studio for students to gain hands-on training? Can you opt for a double major? Does the department have an established internship program? Does the school have a job placement service?

The American Film Institute Guide to College Courses in Film and Television is a good book to assist you in your initial search. (Check "Resources to Locate Training Centers," at the end of this chapter, for ordering details.)

If you are unable to undertake a four-year course of study, consider a two-year program. Some junior colleges and community colleges offer a two-year program in broadcasting. The 69 two-year schools that responded to the BEA survey reported 2,567 first-year students and 2,061 second-year students. These schools provide you with the basic knowledge required for many entry-level jobs in broadcasting.

At the Borough of Manhattan Community College (BMCC) in New York, students train in a million-dollar media center on state-of-the-art equipment. Upon successful completion of a 65-credit program in corporate and cable communications, students are awarded the Associate in Applied Science (A.A.S.) degree.

According to Dr. Sandra Poster, associate dean of academic affairs, BMCC tries to provide excellent hands-on courses, but also requires students to take courses in English, social science, mathematics, physics, art, design, and business organization, and management. "We think this combination provides the kind of education that makes our graduates employable and gives them the potential for upward mobility in a variety of career paths," Dr. Poster said.

Several universities offer courses of study leading to the master's and doctoral degree programs. If you are interested in an advanced degree program, contact the BEA for a list of institutions. The BEA study identified 91 schools with a total of 2,047 students preparing for the master's degree, and 26 universities with 298 students studying for the doctoral degree.

TECHNICAL TRAINING

If you are interested in a technical career in television, you should take a heavy dose of math courses in high school and college. The recommended preparation is a two-year program with an emphasis on electrical or electronics technology, leading to an Associate in Applied Science degree. Then find a job working at a small TV station. While you're working, go to school for a four-year degree. The idea is to get on-the-job experience early. But if you really want to be an engineer with a specialization in telecommunications, you should work towards an electrical engineering degree.

More than 1,200 junior colleges, community colleges, and technical schools prepare students for technical positions in the television industry. For example, Mercer County Community College in Trenton, New Jersey, offers a two-year A.A.S. degree in telecommunications technology. Students are taught to design and lay out plans for TV facilities and maintain microwave and satellite systems.

Several technical institutes offer courses in video maintenance and repair for technically inclined individuals, preparing them for jobs as bench technicians. The Video Technical Institute (VTI) has two centers: one in Long Beach, California, and the other in Dallas, Texas. VTI does not offer single courses, but conducts a 14-month program leading to an A.A.S. degree. Classes start every 10 weeks. The first four semesters are in basic electronics. Other courses include trouble-shooting and video production. According to VTI, the California-based institute is well-connected with the Hollywood and Los Angeles production community. Its average annual student placement is said to be as high as 94 percent.

The De Vry Institute of Technology headquartered in Evanston, Illinois, has campuses in 11 cities in the United States and Canada. Many of them are four-year institutions, but several offer shorter courses for electronics engineers and technicians.

A number of training centers prepare technical students for the FCC license examination. Among them are:

Omega School of Communications
444 North Lake Shore Drive
Chicago, IL 60611

Columbia School of Broadcasting
6290 Sunset Boulevard
Hollywood, CA 90028

Brown Institute
3123 East Lake Street
Minneapolis, MN 55406.

The Electronic Industries Association (EIA) in Washington, DC, offers training programs in basic video and TV repair for technical careers in consumer electronics.

INTERNSHIPS

One of the best ways for students to gain hands-on experience in the field of television is through an internship program. Educators view internships as a bridge between the fundamentals learned at school and the actual demands of the job environment. Employers view internships as opportunities to develop new talent and a prescreening time for new employees. For the student, the time spent learning while working in a professional environment provides an opportunity to build confidence, earn college credits and often even a stipend, and make contacts.

According to Dr. Alan Richardson, associate professor in the Department of Telecommunications at Ball State University, Indiana, "Our university considers the internship program one of the most valuable learning experiences possible during a student's college career." He identifies five primary benefits to interns: 1) They learn to relate to the business environment and understand the performance criteria that are expected from an employee; 2) They learn to relate to their peers on a professional level—often, for the first time in 16 years (interns are not treated as students in the work environment); 3) They make contacts, sometimes leading to a job upon graduation; 4) They are able to add a valuable reference to their résumé; and 5) Often, during an internship, students are able to identify the jobs or environments that are not suited to them.

Several universities have well-established internship programs that provide placement and faculty supervision. If your school does have such a program, be sure to take advantage of it. Frequently upon graduation, students land a job at the station or department where they interned. If your school does not have an established internship program, take the initiative to contact your local cable or television station and ask if you can intern there. You may be able to set up an internship program for yourself by contacting organizations that sponsor interns.

College Broadcaster, the magazine of the National Association of College Broadcasters, is a good source of information on internships available at television stations, corporations and institutions. The magazine is published four times a year; subscriptions are available through NACB membership only.

The Academy of Television Arts and Sciences sponsors 16 expense-paid summer internships to different disciplines for full-time college students. Write to the academy at 3500 Olive Avenue, Suite 700; Burbank, CA 91505. Boston Film/Video Foundation offers internships in film, audio and video. The foundation's address is 1126 Boylston, Boston, MA 02215. Cox Cable Communications sponsors internships in all areas of television production. Write to Cox at 1175 North Cuyamaca, El Cajon, CA 92020.

National Public Radio offers fall, spring, and summer internships at its headquarters in Washington, DC. Positions available in many departments, including news, marketing, and engineering. Candidates must be juniors, seniors, or graduate students, and must be willing to work between 16 and 40 hours per week for an 8- to 12-week period. Academic credit is possible.

If you are interested in cable television, contact the Cable Television Administration and Marketing Society; 635 Slaters Ln., Ste. 250, Alexandria, VA 23314. Internships in cable television are also provided by Turner Broadcasting; 1050 Techwood Drive; Atlanta, GA 30318. Eight internships in cable production serving local educational institutions are sponsored by South Western Cable TV; 8949 Ware Court; San Diego, CA 92121.

You can get additional information on internships by referring to the National Directory of Internships, available from the National Society for Internships and Experiential Education; 122 St. Mary's Street; Raleigh, NC 27605.

NONDEGREE PROGRAMS

All major cities and most smaller cities offer an abundance of opportunities to train for video and television jobs. The popular fascination people have for video, the rapid growth of employment opportunities, and the increased availability of television equipment at affordable prices have led to the proliferation of short courses and workshops. So high has been the demand that several industry veterans as well as some mid-career professionals have gotten into the "seminar business."

Nondegree programs can offer you several advantages. Many of them are conducted on days and at times that meet the scheduling convenience of working individuals. They are usually taught by working professionals, and hence tend to be practical in nature. There are workshops on almost every area of skill specialization and seminars on topics that range from basic issues to future technologies, so you can select from a wide range of options the courses you need to improve your job prospects.

Before selecting a seminar or workshop, get as much information as possible from the sponsor. You will need to know: who the instructor is and what his or her credentials are; what the course content is and perhaps even an outline of topics or skills to be covered; if there are any prerequisites to enrollment; who the course is designed for; and what the course format is (lecture, demonstration, hands-on, field trips). You will also need general information, such as course fee, registration dates, schedule and location, and what the sponsor's cancellation policy is.

There are several types of nondegree programs which include: 1) product-specific training, usually sponsored by manufacturers and vendors of video hardware, software, and services; 2) specific skills training, usually offered by specialists in a specific area of production or by academic institutions that have set up special programs in a specific skill area; 3) general video production training, offered on a year-round basis by nonprofit organizations and commercial enterprises; 4) workshops and seminars conducted on an annual basis in conjunction with annual conventions and trade exhibits. To get you started on your research of nondegree video training, comprehensive information on selected workshops and seminars in each category is presented in the following pages.

Product-Specific Training

The training provided by manufacturers and vendors of hardware, software, or service facilities is usually product-specific. Here is a list of some of the organizations that offer machine-specific training to their clients and those interested in learning how to use specific production hardware or software.

Ampex Training Department
Ampex Systems Corporation
401 Broadway MS 2-11
Redwood City, CA 94063-3199
(415) 367-3702

Avid Technology, Inc.
One Metropolitan Park West
Tewksbury, MA 01876
(800) 949-AVID

Pesa Chyron/CMX Training Department
2230 Martin Avenue
Santa Clara, CA 95050
(408) 988-2000

Grass Valley Training Center
Grass Valley Group, Inc.
400 Providence Mine Road
Nevada City, CA 95959
(916) 478-3000

Weynand Training International
22048 Sherman Way, Suite 212
Conago Park, CA 91303
(818) 992-4481

Weynand offers classes on specific equipment from different manufacturers, such as the Ampex ADO, or the Grass Valley 200 and 300 switchers and 141 editing system, the Quantel Paint Box and Harry, and the Chyron Scribe.

If you are interested in a career as a videotape editor, be it in a staff position or as a free-lancer, you may need some machine-specific training. Write or call the organizations listed above and ask for detailed information regarding cost, location, and schedule.

If you are interested in computer graphics and electronic paint systems, contact the manufacturers of those systems (such as Grass Valley and Silicon Graphics) and ask for details regarding training on their systems.

Specific Skills Training

If you want to improve a specific skill, such as script writing, or specialize in an area such as designing videodiscs, you should look for workshops that offer intensive training in the area of your interest.

The New York chapter of the National Academy of Television Arts and Sciences conducts workshops in television script writing, acting, and directing. You can contact them at 1560 Broadway, Suite 503; New York, NY 10036.

The Nebraska Videodisc Design/Production Group conducts workshops in videodisc design and production. They can be contacted at KUON-TV, University of Nebraska-Lincoln; P.O. Box 83111; Lincoln, NE 68501.

For workshops in computer graphics, contact the Computer Arts Institute in San Francisco for program details, especially, computer graphics in broadcasting. The program is described later in this chapter.

General Production Skills Training

Several organizations offer lectures and workshops on a year-round basis. Some programs are conducted with collaboration from an academic institution and offer Continuing Education Units (CEUs); some programs receive collaboration from unions and associations.

Here is a list of some organizations that offer courses on a variety of video production topics. Use it as a starting point to develop a list of nondegree programs in your area. As you proceed in your research, check the teleproduction and post-production companies in your area for lectures or demonstrations on new equipment or seminars they may sponsor on an ad hoc basis.

American Film Institute (AFI)
2021 North Western Avenue
Los Angeles, CA 90027
(800) 999-4AFI

The American Film Institute provides a broad spectrum of courses in television, video, film, and new media arts and business. The 300-plus offerings range from one-day ($20) to nine-week "Extension" classes in specific skill areas, to a two-year Master of Fine Arts program. Specific courses cover writing, acting, producing, techniques of film, video, computer graphics, animation, and the business aspects of production. The Advanced Technology Program (with the participation of Apple and Sony) enables full time and extension students to become fully literate in the new media, ranging from desktop multimedia to HDTV. Other activities include a grant program for independent video and filmmakers, an annual video contest, and a Television Writers Workshop.

Bay Area Video Coalition (BAVC)
1111 Seventeenth Street
San Francisco, CA 94107

BAVC workshops are held in participating production or post-production facilities and soundstages in the Bay area. Forty courses are offered in each of three semesters a year. Courses are classified as technical, preproduction, production, editing, business, and video technology. Production courses are hands-on, hence are limited to eight students per class. Technical workshops are limited to four students per class.

BAVC courses are recognized and offered as part of The New College curriculum for those seeking academic credit. The courses reportedly are designed especially for free-lance or independent producers as well as the production community at advertising agencies, TV stations, and corporations. Course fees range from $20 for a three-hour seminar to $330 for a four-week production course. An advantage of registering with BAVC is subsidized rates at participating facilities for workshop students who have completed 11 hours or more of coursework.

Center for Creative Imaging
51 Mechanic Street
Camden, ME 04843

The center offers courses in creative imaging, graphic animation, desktop video and desktop music production. Courses are conducted year round.

Computer Arts Institute (CAI)
310 Townsend Street, Suite 230
San Francisco, CA 94107

CAI course fees range from $165 for a three-week storyboarding course to $3,132 for a four-month foundation course in 2-D and 3-D computer graphics and computer design for broadcast courses. Most CAI courses are held at its own facility, but some sessions are held off-site. The three-semesters-per-year program includes 39 titles in computer graphics, 2-D and 3-D machine courses utilizing the computer laboratory, and five courses in computer design for broadcast. A certificate is awarded upon completion of the four-month foundation career curriculum and the advanced career curriculum, respectively. Off-site, hands-on sessions, such as the Quantel Paint Box workshop, are limited to eight students per session.

Dynamic Graphics Educational Foundation (DGEF)
6000 North Forest Park Drive
P.O. Box 1901
Peoria, IL 61656-1901

Equipment-intensive workshops are offered only at the DGEF Training Center in Peoria. Other workshops are offered at hotel sites in several cities. DGEF conducts two semesters annually, the first from March through July and the second from August through December. Three video workshops are offered each semester and each workshop is usually a five-day program. The hands-on video and computer graphics production classes are limited to 20 students per course. Fees for a video course are approximately $895. You can earn 3.5 CEUs for a five-day video workshop conducted by DGEF.

Film Arts Foundation (FAF)
346 Ninth Street, 2nd Floor
San Francisco, CA 94103

FAF offers approximately 13 titles a year. Only two or three are specifically in video; the majority are core technical courses in film production, and a business session is offered once a week as part of an 8-to-12-week business series. Fees range from $15 to $25 for an evening class and from $40 to $60 for a weekend course. FAF has a 16mm film facility. Hands-on courses are limited to 8 to 15 students, depending on the course. FAF members receive a monthly newsletter listing scheduled classes. If you are not a member but would like to receive seminar and workshop information on a regular basis, write to FAF and ask to have your name and address put on the mailing list.

Half-Inch Video
185 Berry Street, Suite 467
San Francisco, CA 94107

Half-Inch Video is a production and post-production facility. Approximately five times a year, it offers a public video seminar free of charge. The seminar is usually an hour's duration. The company also conducts three-hour private courses on an on-demand basis. The private courses are conducted at the facility for a $75 fee. Two types of courses are offered: 1) VHS editing, and 2) the use of a character generator. These courses offer one-on-one, hands-on instruction, so class size is limited to four students.

International Film & TV Workshop
2 Central Street
Rockport, ME 04856

The International Film and Television Workshops operate from May to October in a rustic New England setting. More than 100 hands-on training sessions are offered on subjects relating to the craft of film and video—such as directing, shooting, editing, animation, and cinematography. The workshops range from one week to the entire summer, depending on the area of study. Most one-week workshops range in cost from $700 to $900, not including meals and housing.

The Sony Institute of Applied Video Technology
2021 North Western Avenue
P.O. Box 29906
Los Angeles, CA 90029

The Sony Institute, now in its twenty-third year, offers workshops at its headquarters in Hollywood as well as at other locations all across the nation, including university facilities and Sony video dealer facilities. Thirty different workshop titles are offered nationwide, nearly 50 percent are held in California. Workshops vary from two to five days in duration. The fee ranges from $599 to $1699 per course. A diploma is awarded on the completion of every workshop. You can call Sony and request to be put on the mailing list for up-to-date workshop information.

Annual Seminars And Workshops Held In Conjunction With Conventions And Exhibitions

The majority of seminar programs held in conjunction with annual conferences and trade shows are sponsored by professional associations. Some are sponsored by commercial enterprises. The advantage of attending such workshops and seminars is that they frequently are conducted by the best people in the industry. Most pros consider it an honor to speak or lecture at an association's annual conference and hence are willing to travel out of their hometowns to present a paper or conduct a workshop. This is the best way to hear several renowned speakers or experts at one location. Another advantage of attending an annual seminar program is that in most cases you also can attend the trade exhibition of the latest equipment and software.

Chapter 9 discusses annual conferences and conventions as well as professional associations. Make your own list of the professional associations and annual conferences that will best meet your professional needs. Directories listed in Appendix A will help you identify associations you may wish to contact. Using Appendix B, write or call the association and ask if they offer a seminar program in conjunction with their annual meeting. Also ask if an exhibition or trade show is held at the annual meeting. Some associations host an extensive workshop program at their annual meeting but do not host a trade show. The ITVA is one such association.

In addition to association-sponsored annual seminar programs, you should consider attending the workshops offered by publishing companies that sponsor annual trade shows and exhibitions.

At Image World—held annually in New York, Los Angeles, San Francisco, Orlando, and Chicago—an extensive seminar program is offered. Contact Knowledge Industries Publications, Inc. (KIPI); 701 Westchester Avenue; White Plains, NY 10604; or call the company in New York at (914) 328-9157.

In conjunction with the annual Optical Information Systems conference, Meckler Corporation sponsors a seminar program. Contact the company at 11 Ferry Lane West; Westport, CT 06880; or call (203) 226-6967.

RESOURCES TO LOCATE TRAINING CENTERS

Information on courses given at various schools is available from the Broadcast Education Association (see Appendix B for address), which was founded in 1955 to develop a closer working relationship between the academic world and the world of professional broadcasting.

A directory which briefly describes cable training schools can be purchased for $1 from the National Cable Television Association's Publication Department (see Appendix B for address). The booklet is organized by states, making it easy to find a program in any specific location.

The American Film Institute Guide to College Courses in Film and Television is published every two years and can be ordered by mail from Book Order Department; Peterson's Guides; P.O. Box 2123; Princeton, NJ 08543-2123. The AFI guide is a comprehensive listing of institutions that offer a program for film and television studies. Listings are classified by type of institutions: two-year institutions; four-year institutions; upper-level institutions that begin with the junior year and award bachelor's degrees and offer graduate work; professional institutions that offer graduate work; professional institutions that offer programs of study in one field only, leading to an undergraduate or graduate degree; and universities which offer four years of undergraduate work, plus graduate work through the doctorate, in more than two research-oriented and professional fields.

Information about colleges and universities offering courses in radio, television, and film can be obtained from the Speech

Communication Association; 5105 Backlick Road; Annandale, VA 22033.

If you are interested in training for professional video work in nonbroadcast industries or institutions, contact the International Television Association (ITVA) for a list of professional video programs in higher education. The list is the result of a 1985 national survey conducted for ITVA by Michael J. Porter, Ph.D., and Barton L. Griffith, Ph.D., at the Department of Speech and Dramatic Art, University of Missouri-Columbia.

The National Directory of Internships can be obtained for a fee from the National Society for Internships and Experiential Education; 122 St. Mary's Street; Raleigh, NC 27605.

For a list of junior colleges and community colleges, contact the American Association of Community and Junior Colleges (AACJC), Washington, DC.

CHAPTER 9

PROFESSIONAL DEVELOPMENT

Landing your first job in the television industry may not be an easy task. More difficult, however, are holding onto that job, making the right career moves, and building a successful and rewarding professional career. But once you have decided to make a start, your professional development should be a lifelong affair. The intensity of your desire to improve will determine how much time and commitment you invest in continuing media-related education and participation in professional associations and other industry events. All this in addition to a full work load on the job!

The ongoing task of professional development is more easily accomplished by participation in festivals and competitions, conferences and conventions, and professional associations. In addition, there are many books, directories, trade magazines, and newspapers published in this industry to keep professionals up-to-date and aware of trends. Reading is an essential activity in professional development.

In the next few pages you will find information on each of the components of professional development. Some insights are provided so as to give you a better understanding of the benefits of planned professional development. You also will find a list of selected festival names and sponsors, major conferences and their sponsors, professional associations, and periodicals. None of the lists are exhaustive, but they are comprehensive enough to give you a good start in your research for resources. Use them to create your own professional development strategy. For information on media-related continuing education seminars and courses, check Chapter 8.

FESTIVALS, AWARDS, AND HONORS

The glamor and excitement at an awards ceremony in the television industry is truly electrifying. You may have caught a little of the excitement just by watching the Emmy Awards on television. The red plush seats and crystal chandeliers, the powerful searchlights and giant screen, the evening gowns and tuxedos—all add up to that heady feeling of success and stardom.

Many organizations bestow recognition on their members for the outstanding quality of their work. Some honor excellence in the craft areas of the video industry; others recognize technical achievements, including innovations; some present awards to individuals who have contributed significantly to the advancement of the utilization of video. Several associations honor the distinguished service of those individuals whose contributions of leadership and dedication to the industry extend above and beyond the call of elected office.

One way you can build your credentials is by entering and winning awards at television, video, and film competitions. An award is a definite plus on your résumé and can give you a major career boost. It will set you apart as a top-notch producer-director, camera operator, editor, or whatever your area of excellence may be.

When a program wins an award, it does wonders not only for the individual on whom it is bestowed but also for the production crew. In the broadcast industry it spotlights the station. In the nonbroadcast industry it brings stature to the communications department. Award winners agree that winning an award is a great morale booster and that it elicits approval and support from top management, be it in broadcast or organizational television.

Television and video festivals have a way of furthering the standards of achievement. Each year's festival entries appear to raise the level of excellence. Participating in festivals can only reinforce your confidence and spur you on to greater heights of achievement. Video festivals give participants a tremendous sense of pride and accomplishment.

Walter Hamilton, national chairperson of the 1986 Monitor Awards, likened awards gala events to a harvest festival: "At the end of the harvest, when all the crops had been gathered and the flocks and herds had been fattened, the people came together to celebrate. The toil and all the hard work were behind them, and it was time to praise the Lord and make merry with friends, old and new. To stop for a day and let the cares of the past year fade away in song and dance and laughter.

"We are urban folk, and the fields we plow are different, but the toil and the hard work remain. The struggle is still there. We come together to bask in the joy of the event. Some people will be winners and some people will be nonwinners, but most of the people there will not be involved in winning or losing. They are there to have a good time with friends and colleagues."[1]

Most trade journals and magazines carry information on video festivals and competitions. If you intend to participate, it is best to contact the sponsor early for detailed information and entry-application forms. There are a great number of entry categories which include commercials, public service announcements, and music videos in broadcast. Nonbroadcast program categories include employee communications, training, sales or marketing,

---

[1]Walter N. Hamilton, "Why the Monitor Awards?", *1986 Monitor Awards Program,* International Teleproduction Society, Inc., p. 18.

and information. In some competitions, student entries are judged in a separate category.

A number of well-established national competitions and festivals are listed here. For details on those sponsored by professional associations, contact the sponsor at the address and telephone number listed in Appendix B. The address and telephone number of those sponsored by commercial enterprises are listed here.

The International Television Association sponsors an annual video festival designed to honor individuals whose work contributes to the advancement of excellence in professional video communication. The entry deadline is usually in November. The entry fee is based on member, nonmember, student, or student-nonmember classifications. Entry tapes must be submitted on the 3/4-inch NTSC U-matic format, or 1/2-inch NTSC, VHS format with identical audio mixed on both channels.

At its annual convention, the ITVA presents the Golden Reel™ and Silver Reel™ awards for programs that effectively achieve their communication objectives and demonstrate very high standards in video production, technical quality, and creativity. At its 25th annual conference and awards ceremony held in Phoenix, Arizona in 1993 the association presented 11 Golden Reels and 22 Silver Reels. Winning videos were chosen from nearly 800 entries.

The association also presented one special achievement award, and one student achievement award. Entries are accepted in the following categories: training, sales/marketing, internal communications, external communications, organizational news, interactive video, public service announcements, videoconferencing, student productions.

The International Teleproduction Society sponsors the Monitor Awards at an annual awards banquet. Among its 17 entry categories are cable entertainment, children's programming, sports, local and national commercials, news or documentaries, special effects, and audio for video. The entry deadline is usually near the end of January. Separate entry fees apply to members and nonmembers.

The New York Festivals hosts the international Film and TV Festival of New York. It sponsors three separate competitions, one for commercials, another for nonbroadcast programs, and a third for TV programs, promotional spots, and music videos. The entry deadline is usually in September. The fee varies according to entry.

Commercials are accepted on 3/4-inch U, 16mm or 35mm. Nonbroadcast industrial or educational programs are accepted only on U-format videotape or 16mm film. Filmstrips or slide presentations must be on 16mm or 35mm. Multimedia or multi-image programs are accepted only on 3/4-inch videotape. TV programs, promotional spots, or music videos must be on 3/4-inch videotape or 16mm film.

For festival information, contact the sponsor at 655 Avenue of the Americas; New York, NY 10010; or call (914) 238-4481.

The American Film and Video Association sponsors the American Film and Video Festival. The entry fee varies according to the length of program. The entry deadline is usually mid-January. Programs are accepted on 3/4-inch videotape or 16mm film.

The Golden Quill Awards are presented by the International Association of Business Communicators. The entry deadline is in January. The fee is $40 for members and $80 for nonmembers. Programs are accepted on 1/2-inch videotape, film, and slides.

The National Federation of Local Cable Programmers, in conjunction with Fuji Photo Film U.S.A., sponsors the annual Hometown USA Video Festival. The entry deadline is mid-March. The fee ranges from $20 to $30. There are 31 categories. Entries may be submitted on 3/4-inch videotape or both 1/2-inch formats (VHS and Beta).

The International Computer Animation Competition is sponsored by the National Computer Graphics Association. The entry deadline is usually in November. The fee varies according to professional or nonprofessional status. There are many categories, and programs may be submitted on 3/4-inch NTSC or PAL or 1/2-inch VHS.

Image Film and Video Center hosts the Atlanta Film and Video Festival. The deadline is December. Videotape entries are accepted in NTSC only, on 1/2-inch VHS or Beta, 3/4-inch U, and Super-8 or 16mm film. There are many categories. For details, contact the sponsor at 75 Benett Street, Suite M-1; Atlanta, GA 30309.

The Telly Awards was founded in 1980, to showcase and give recognition to outstanding nonnetwork and cable commercials. The competition was expanded several years ago to include film and video productions. In 1992, the competition drew 5,600 entries in all categories, including 2,500 in nonbroadcast film and video. Entry forms may be obtained from Telly Awards, 4100 Executive Park Drive, Cincinnati, OH 45241.

Women in Communications, Inc. sponsors the National Clarion awards competition, which recognizes the best in communications in 85 categories including television. For more information contact the sponsor at 2101 Wilson Blvd., Ste. 417, Arlington, VA 22201.

The Association for Multi-Image hosts the International Multi-Image Festival. The entry fee is $145 for members and $225 for nonmembers. The entry deadline is usually near the end of May. Programs are accepted on 3/4-inch U-format NTSC only.

The Cindy competition is held by the Association of Visual Communicators. The entry deadline is in June. Programs are accepted in several formats, including videodisc. Programs are judged in many categories.

The Council on International Nontheatrical Events hosts the CINE competition. Entries are accepted on 3/4-inch videotape and 16mm film.

Movies on a Shoestring, Inc., sponsors the International Amateur Film Festival. The entry fee is $7. Entries are accepted on 8mm, Super-8, 16mm, 3/4-inch, Beta, and VHS. For details write to MOAS; P.O. Box 17746; Rochester, NY 14617.

The Retirement Research Foundation sponsors its National Media Awards. Programs on aging issues are accepted on videotape and film. For more information write to CNTV, 11 East Hubbard Street, 5th Floor; Chicago, IL 60611.

In addition to national festivals and competitions, the regional and local chapters of several associations sponsor their own awards competitions in order to recognize accomplishments in individual community settings.

The ITVA Chicago chapter sponsors an award for "bloopers and blunders" in a delightfully embarrassing evening themed "The Turkey Shoot." Golden and Silver Gobblers are awarded in two categories: intentionally funny and unintentionally funny. This competition has brought to screen a priceless commodity in the art of television production: the outtake.

"We have got it all on tape: the tongue-tied CEO, the 27th take when the whole crew gets the giggles—the exquisite moments in TV production that are meant to be shared, not hidden away in stock footage vaults," said Bonnie Farnon of the ITVA Chicago chapter. According to Farnon, members see the event as a reward in itself for all the hard work throughout the year, and "a great opportunity to laugh at our own mistakes—and everybody else's."

CONFERENCES AND CONVENTIONS

Whether you are a beginner or an "old pro" involved in any aspect of the television industry, your attendance at a trade conference at least once a year is a *must*.

Television conventions and exhibitions provide you with an opportunity to try out the latest equipment for creating superb programs. They offer you a shopping arena for the products and services you need. On the exhibit floor, leading suppliers display their new product introductions and let you experiment with demo equipment. You can talk with product representatives about your needs or get technical advice and solutions to your production problems. Attending a trade show is one way of keeping tabs on technology.

Several associations offer a seminar program at their annual conference. At a conference or seminar you will have the opportunity to expand your knowledge of proven techniques for television production, distribution, and management. Professional seminars are offered at all levels of expertise, from basic to advanced. Full- and half-day seminars are offered on a wide variety of topics, broadly classified as follows: production, post-production, audio-forevideo, technical, management. General sessions include topics on leading-edge technology as well as such practical topics as "Jobs in Video." Seminar speakers are high-caliber, awarding-winning professionals with a wealth of experience to share. What you can learn from their insights is often much more than what classrooms and textbooks can offer.

Registration fees for conference attendance vary according to the sponsor and whether or not you have membership status with the sponsoring association. Some commercial enterprises sponsor a trade event which includes a seminar program and exhibition. At some of these events, your registration fee for a seminar will entitle you to free admission to the exhibition. Fees for seminars can cost you a couple of hundred dollars. Admission fees to an exhibition are about $20 per day. Exhibitors at trade shows receive a good number of free passes to the exhibition, which they give to their preferred clients and potential customers. If you need a free pass, contact your local representative of an exhibitor company early, because there is a great demand for these passes at major shows.

If you are a full-time employee, your company may pay for your attendance at a professional conference once a year. Many corporations have an annual budget allocation for continuing professional education or training. Several video managers at large corporations report an annual departmental budget of over $5,000 for the continuing professional training of their staff. If you are allowed only one conference a year, you will want to select very carefully the one you will attend. You should identify your objectives for conference attendance clearly so that you will get the maximum benefit from it.

Another advantage of attending seminars and conferences is the opportunity to meet and exchange views with colleagues who work in different parts of the country. You will undoubtedly expand your network of industry contacts at these meetings. Talk to as many people as you can. Let them know if you are looking for a job, and be specific as to what kind of job you want. You will be surprised at the number of job opportunities that are made known and filled at a conference.

Some conferences have a job posting or referral service on-site. Be sure to check this opportunity for finding employment. Call the association and ask if the service is available. A good way to be prepared for such an opportunity is to bring your résumé and a tape of selections of your work to the conference. Bulletin boards for "positions wanted" cards are on display at some conventions. Post your name, address, and phone number.

A major industry event is the NAB Annual Convention and International Exposition hosted by the National Association of Broadcasters. Billed as the world's showcase of broadcast equipment, it attracts nearly 40,000 professionals to the exhibit booths of over 600 manufacturers and vendors of broadcast and professional-grade products and services.

The following selected list consists of associations that sponsor annual conventions and conferences:

AMI, Association for Multi-Image. International convention offers an exhibition of equipment and services as well as showings of finalists in the annual competition.

AES, Audio Engineering Society. Convention, presentation of papers, workshops, and equipment exhibits.

ICIA, International Communications Industries Association. INFOCOMM International exhibits of video, computer graphics, teaching and instructional equipment, and material. General sessions and seminars.

ITCA, International Teleconferencing Association. Meetings and exposition of exhibits and seminars (INTELEMART).

ITVA, International Television Association. Conference seminars, meetings of special interest groups, showings of video festival entries and awards. No exhibition of equipment.

NAB, National Association of Broadcasters. Convention, exhibits, and conference concerning television and radio.

NACB, National Association of College Broadcasters. National Conference in Providence, RI. Panel discussion, seminars, exhibits, and job fair.

NATPE, National Association of Television Program Executives. Conference of workshops, meetings, and programming marketplace. Separate production conference on television draws production people and equipment manufacturers.

NCGA, National Computer Graphics Association. Exposition of products, systems, and services. Educational programs featured.

NRB, National Religious Broadcasters. One annual and six regional conventions.

RTNDA, Radio-Television News Directors Association. Convention and conference.

SALT, Society for Applied Learning Technology. Conference on interactive technology and product exhibits.

SIGGRAPH, Special Interest Group of the Association of Computing Machinery. Conference includes technical presentations, tutorials and seminars, a large exhibition of hardware and software products, and an art exhibit.

SMPTE, Society of Motion Picture and Television Engineers. Fall convention offers papers and equipment exhibition. Winter gathering features papers on a few main topics. Equipment exhibited is directly related to papers.

Image World. Regional equipment exhibits and seminars for those actively engaged in television production and distribution for nonbroadcast organizations. Knowledge Industry Publications Inc.; 701 Westchester Ave.; White Plains, NY 10604.

PROFESSIONAL ASSOCIATIONS

The best way to enhance your career in an industry is to join a professional association. These organizations, consisting of groups of individuals with common career objectives and vocational interests, are set up primarily as a communications vehicle for members. An association serves as a forum for the exchange of ideas.

Television professionals depend on their associations to provide an environment for the sharing of ideas and information about emerging technologies and their applications. When you join a professional association, you become part of a support system that provides access to a wealth of resources. The most important benefit is the chance to meet and interact with other professionals. People working in television and video agree that the one thing most precious to a career in the industry is contacts.

Kathy Morris, currently president of Chicago-based Morris Communications and member of the ITVA board of directors, states the case for "networking" succinctly:

> Video communications is not an industry of "one-man bands." Oh, it may seem that way at times . . . like when you're slumped over a script at 4:00 a.m. . . . or on Saturday afternoon when you're recutting the 15-second, 30-edit montage open because you caught a glitch during the final screening that no one else noticed . . . or when your boss' offhand comments during a staff meeting about equipment budgets becomes, "Let me have the dollar figures and rationale for the new studio by next Friday."
>
> It is easy to feel that the weight of the production, the department, the whole company rests on your shoulders. Thank God for contacts.
>
> It is during moments like those that the value of colleagues becomes clear . . . when the three writers you met at the ITVA conference and who sent you their ["telephone techniques"] sample scripts get you past your "writer's block". . . or when you find the home phone number of the engineer at the local post-production facility who patiently talks you through an adjustment to your R VTR so you can have the master to the duplicator first thing Monday . . . or when you meet the manager of another company's video department who, over breakfast, outlines the exact equipment package you need to get started and tells you whom to call for current prices.
>
> Need a job? You need contacts. Need the best 3/4-inch VCR maintenance person? You need contacts. Need 30 seconds of the Golden Gate bridge and the Statue of Liberty? You need contacts.
>
> Contacts mean you can get the job done, no matter what the job is, even if you are the only person on the video staff. Contacts mean that while your facility may not be "edge-of-the-art," you know whose is and how to take advantage of that. Contacts mean that manufacturers and distributors

> of equipment are people you rely on for information and whose insights you respect.
>
> Professional associations make it easy to make contacts, and contacts make the video communications industry an exciting place to be.[2]

Your membership dues will buy you more than just an opportunity to make contacts and network within the industry. Association benefits include job lines, salary surveys, membership directories and other publications, and seminars and conferences. Salary surveys conducted by associations enlighten members as to appropriate compensation. Job descriptions enable them to clarify their function and responsibilities to company management.

Several employers have a company-sponsored membership program through which their company will pay an employee's membership dues and related expenses for meetings. Employers expect their staff to play an active role in professional associations. Several corporations encourage their employees to serve on committees and play a leadership role in the industry.

Listed below are some professional associations, with brief descriptions of their activities and the specialized groups they serve within the television and video community. Addresses of these organizations appear in Appendix B.

Most of the associations listed below are international in character and accept members worldwide. Many groups have chapters and international affiliates, and some offer student memberships. Benefits usually include a newsletter, magazine, or journal, plus group medical insurance and discounts on meeting fees.

---

[2]Kathy Morris, "No Man is a One-Man Band," *International Television, The Journal of the International Television Association.* May 1984. p. 4.

AES, Audio Engineering Society. Members work professionally in audio or have an engineering degree in audio. Papers and workshops are presented at section meetings and annual conventions. Student membership is available.

AMI, Association for Multi-Image International. Members are actively engaged in multi-image production and utilization. AMI promotes the use of multi-image as a medium for education, communication, and entertainment. Members include communication directors, computer graphics designers, audio engineers, scriptwriters, and television production people. Chapters and international affiliates hold regular meetings. Special student rate.

ICIA, International Communications Industries Association. Members are producers, manufacturers, and dealers of equipment and software used in audio and audiovisual communications by business, industry, government, education, and health care. ICIA offers a certification program for business and sales professionals in audiovisual equipment, software, and services. The association is active in legislation matters affecting AV manufacturers and dealers.

IICS, International Interactive Communications Society. Members are engaged in the production of interactive communications programs. The society promotes the use of interactive communications in business, industry, education, and the arts. Activities include regular chapter meetings, a shared disc program, and special programs and workshops.

ITCA, International Teleconferencing Association. Members are users, researchers, and providers in the field of teleconferencing. ITCA advocates and promotes research, applications, and development of improved teleconferencing systems and services. The association addresses issues involving teleconferencing before local and national regulatory and legislative bodies.

ITS, International Teleproduction Society. Members are producers and manufacturers in the teleproduction industry. Members are engaged in the business of video and/or audio production, post-production, videotape duplication, film-to-tape transfers, and standards conversion.

ITVA, International Television Association. An association of professionals producing video in nonbroadcast organizations. Members are managers, producers, and production people. Members represent a variety of industries, categorized by special interest groups (SIGs). There are ITVA chapters worldwide; many hold monthly meetings. Student memberships and scholarships are offered.

NAB, National Association of Broadcasters. Members are television and radio executives, managers and engineers. Since its founding, NAB has represented the broadcast industry in discussions and negotiations with the FCC and Congress. It works with other industry groups to affect policies and laws in a positive way.

NATPE, National Association of Television Program Executives. Members are station program managers and other professionals. NATPE provides faculty development grants.

NCGA, National Computer Graphics Association. Members need only to have an interest in computer graphics. NCGA promotes computer graphics and is devoted to graphics applications in business, industry, government, science, and the arts. It provides a forum for communication between vendors and users and supports standards for computer graphics.

NCTA, National Cable Television Association. Members are U.S. cable TV systems operators, equipment manufacturers, and program suppliers. NCTA lobbies for the U.S. cable television industry.

NRB, National Religious Broadcasters. Members are organizations which produce religious programming for radio and television and operate such stations. The NRB conducts research on audiences, programs, and the effectiveness of religious broadcasting.

RTNDA, Radio-Television News Directors Association. Members are news directors and others involved in electronic journalism. One of RTNDA's principal objectives is the fostering of journalistic freedom to gather and disseminate information to the public. Student membership available.

SALT, Society for Applied Learning Technology. Members are instructional technologists, and education or training program developers. The organization is involved in the development of interactive instructional materials for business, industry, education, and training.

SBE, Society of Broadcast Engineers. Members are radio and television engineers. The SBE certification program confirms that a member engineer has successfully completed certification and is qualified to perform the duties of specific levels of competence. Student membership category.

SIGGRAPH, Special Interest Group on Computer Graphics. A group of the Association of Computing Machinery. Members are artists and technicians who work in production and post-production facilities and for advertising agencies and corporations, equipment manufacturers, and providers of software. SIGGRAPH provides a forum for computer graphics research and for applications.

SMPTE, The Society of Motion Picture and Television Engineers. The society's specialized groups develop and establish standards for film and television technologies as they are used in theatrical entertainment and news and in industry. Local chapters hold regular meetings. Student membership category.

PUBLICATIONS

To keep abreast of developments in this rapidly changing field, you must read trade journals and magazines that address your specific job function. For example, if you are a television engineer, you should read the *SMPTE Journal.* Likewise, if you are the director of a corporate communications department, you should read *AV/Video* or *Video Systems.*

There appear to be as many magazines as there are specializations in this field. Many of them address the issues and concerns in overlapping areas such as video and computer graphics. Ideally, you should read or at least scan as many trade publications as you can. Carefully select the magazines you want to receive. Make sure the ones you subscribe to are related to your job environment and specific job activity. If you are an indiscriminate subscriber, you may waste some money, but worse still, you will be overwhelmed with mail!

Read trade magazines regularly, as they provide articles on innovative applications areas as well as tips on trouble-shooting. Columns in trade magazines are often written by television or video pros who can give the reader practical production and management advice. Feature articles are usually thoroughly researched. Some offer in-depth analysis of outstanding productions and focus on unique teleproduction techniques used in award-winning programs. News sections include briefs concerning people and companies, including changes and accomplishments affecting the industry. Several magazines publish lab test results and critical reviews of new products, indicating their appropriate application areas. Many periodicals include help-wanted sections; some include positions-sought sections as well.

The periodicals listed below are classified on the basis of their editorial coverage. Addresses appear in Appendix A.

Broadcast Communications
- *Broadcast Engineering*
- *Broadcasting & Cable*
- *Television Digest*
- *TV Technology Wrap*

Teleproduction or Post-Production
- *Film & Video*
- *On Location*
- *Millimeter*
- *Mix*

Cable
- *Cable Age*
- *Cable Communications*
- *Cable Marketing*
- *Cable News*
- *CableVision*

Organizational TV
- *AV/Video*
- *T.H.E. Journal*
- *Videography*
- *Video Systems*

Engineering or Technical
- *Broadcast Engineering*
- *IEEE Spectrum*
- *SMPTE Journal*

Multimedia
- *CD-ROM World*
- *CD-ROM Professional*
- *Computer Pictures*
- *Imaging*
- *Mac User*
- *New Media*

Subscriptions to most trade publications are free for those who qualify as professionals. All you have to do is fill out and mail a subscription card, which is available from the publisher and often is bound into the magazine. If you do not qualify for a free subscription, you can receive a magazine by paying its cover price. Appendix A lists trade publications.

Journals published by professional associations usually are mailed to members as a membership benefit and at no cost to the member. Some regional and local chapters of national associations periodically publish a newsletter. These are an invaluable source of information because they frequently list job openings as well as special events in the area.

In addition to media-related literature, you should read selected periodicals related to your job setting. For example, if you produce or manage video for a hospital or health care clinic, you should read industry-related literature such as *HESCA Feedback* in order to be up-to-date. Some special interest groups and committees of professional associations publish newsletters which may better meet your specific job-setting information needs.

CHAPTER 10

THE JOB SEARCH

Professional communicators have frequently been accused of an inability to communicate! The skill most required in looking for a job is the art of communicating effectively. Expressing yourself clearly—both in writing and in speaking—is essential for the two critical steps in landing a job: preparing a résumé and being interviewed.

Basic communications skills should be brought into play when looking for a job. You will need to research the job market, write a résumé and cover letter, and perhaps even produce a demo tape of your video production work. You may need to sell your skills (even by using advertising media) and conduct your own public relations.

While the whole field of communications is growing at a rapid pace, so also is the number of trained people trying to enter the field. The marketplace is truly competitive. Nevertheless, it is possible to get the job you are looking for, if you go about it the right way. Be prepared for hard work and some disappointments, too. Looking for a job can, in itself, be a full-time job.

RESEARCH

The most critical factor in job-hunting is research geared toward targeting a job market. Once you have determined your area of interest and your career goals, you should make a list of specific industries and companies to approach. Each chapter of this book lists the main directory which will provide you with names of employers in specific areas of television or video. For your quick reference, they are also listed in Appendix A. Use them to develop your target list. In addition, you should use the membership rosters of associations in the field to identify the names of the hiring executives. Read trade publications that report on people moves within the industry to update your list.

Your research should also include brief profiles, in particular the needs, goals, and philosophy of the organizations where you seek employment. Many of the job-search books listed in Appendix C include resource sections on how to learn more about the specifics of an industry and a company. In addition, there are several directories of corporations which list company addresses, telephone numbers, and information such as the nature of the business, the company's annual sales, and the names of officers and directors. These directories are available in the reference section of public and university libraries.

NETWORKING

Most jobs in the television and video industry are obtained by word of mouth. That is why it is very important to start making and maintaining contacts with people in the profession at an early stage of your career. Maintaining contacts is an ongoing process.

Creative people often find it difficult to admit that they are looking for a job. But your chances of finding a job will be greater if more people know that you are looking for one. Often, the best way to find a job is through people who know you and what you are capable of doing. So tell your family and friends, and their friends, as well as alumni and other professionals, that you are looking for a job. Be specific—let them know what kind of job you are seeking.

It always helps to meet people who are doing the kind of work you are interested in. They may offer valuable advice on job-hunting and perhaps introduce you to hiring executives. Ask to see people in the industry even if they do not know of a job opening. Use these meetings as informational interviews to learn more about an area of the industry or specifics, such as how a station operates, or about the corporate culture of an in-house TV studio. Such pre-job interviews frequently generate job leads. In any case, it will lead to more contacts and expand your professional network.

JOB POSTING AND HOTLINE SERVICES

The traditional job posting service is in the form of a bulletin board, usually found in a central location at a college or university department of telecommunications or media studies. Most academic institutions continue to post job opportunities in this manner. In addition, the placement offices at a few universities are very active in seeking out employment opportunities and are successful at making a good number of placements a year. Some campus placement offices maintain updated résumés on file as a service to their graduates, at a nominal annual fee. The résumés retained on file are available for employers to review.

Several local professional organizations operate *job banks* or *job lines* where openings are publicized, usually at monthly meetings and through newsletters. For example, the Computer Arts Institute (CAI) in San Francisco operates a job placement service that works in two ways: 1) Job opportunities are posted on the bulletin board. This service is an advantage not only for current students, but also for those who attend events, such as "Video Night," that are open to the public free of charge; 2) A search for candidates to fill specific job openings is conducted on behalf of employers, at a nominal fee charged to the employer. CAI maintains on file résumés of current students, as well as graduates, for employers to review. The second service is of benefit only to CAI students and graduates.

Many national associations offer job placement services at no cost to the job seeker. For example, the National Association of Educational Broadcasters offers assistance to anyone looking for a position in public broadcasting, through their service called PACT. The NAB operates Employment Clearinghouse (ECH), which is a recorded Jobline featuring current nationwide radio and television station jobs sent to ECH weekly. Openings in a particular job category are featured as follows: Monday–on-air talent, Tuesday–sales, Wednesday–production, Thursday–engineering, Friday and Weekends–news.

The ITVA operates Job Hotline as a service to its members. The Hotline reportedly receives an average of 102 calls per week from job hunters and lists 8 to 10 jobs per tape. Job openings are announced at no charge by employers who may or may not be members of ITVA. However, job applicants must be current members to take advantage of the service.

This is how the service operates: job openings are listed on a three- to five-minute audiotape that runs 24 hours a day, seven days a week. The tape is updated approximately every two weeks. The tape begins with instructions to applicants regarding necessary information to be provided as well as a summary of all positions listed. This lets the listener know what is available where, without listening to the entire tape. The hotline then gives detailed information for each position, such as job duties, starting date, and salary. The amount of detail is at the discretion of the employer. All listings are confidential, and are itemized by job code number, with reference only to the job title, city location, and short job description.

To apply for jobs listed on the Job Hotline, a member should mail a résumé and cover letter with job code and ITVA membership number on the front of the sealed envelope to ITVA's headquarters in Irving, Texas. Each position applied for must have a separate envelope. All envelopes with the correct format are then forwarded, unopened, to prospective employers. Employer response is not guaranteed by this service.

JOBS AND CAREER INFORMATION ON-LINE

If you have access to a computer with a modem, you can obtain career related information through the following channels: Commercial Online Services; Electronic Bulletin Board Services; and Internet.

Commercial Online Services: America Online (AOL), CompuServe, and Prodigy are an invaluable source of information on job opportunities. America Online, for example, has a "Career Center" which provides a wide range of job search related information, such as *E-Span* and *Help Wanted-USA* job listings, for basic on-line charges (approximately 10¢ a minute). Numerous special interest groups (SIGs), and electronic forums ranging from educational T.V., to broadcasting and video production are also accessible through these services; as well as through independent "bulletin boards" and the Internet.

BBSs (Bulletin Boards Services): Professional groups, government agencies, special interest groups (SIGs), and thousands of ordinary folks with a hard disk, a modem, and a dream have established their own BBSs, where information, discussions, and downloadable (retrievable) software are available at little or no cost beyond that of a phone call.

One example is the State of Virginia Teleconferencing BBS. It offers news about conferences and meetings, teleconferencing and electronic communications, and hundreds of files of related information (including lists of other BBSs).

Internet: The Internet is the great "data highway" connecting virtually every university computer and E-Mail system, government agency, major corporation, and communication utility in the world.

It is comprised of thousands of separately administered networks, each of these connecting as many as tens of thousands of computers. The total number of individual users of the Internet is in the millions. Online users might be interested in "subscribing" to some of the over 3000 Internet "listservs." Listservs are electronic discussion groups operated on the Internet. They make it possible for participants to discuss topics of professional or personal interest with others who share their interests. For example, the National Association of College Broadcasters (NACB) posts job listings on its *listserv.*

Access to the Internet is available through most institutional and corporate E-Mail systems, Bitnet, commercial services (AOL, Compuserve, Prodigy, etc.), and many other routes. For more information on the Internet, contact your local E-Mail coordinator, computer guru, or *Boardwatch* Magazine.

DEMO TAPE

Traditionally, a demo reel or tape was a requirement mainly for on-camera jobs. However, don't be surprised if you are asked to present a sample tape when applying for a video production job. The relative ease with which employers can access playback equipment today is making demo tapes commonplace. A sample tape can be your key to success. Competitive as the job environment is, it is important that you project your skills, talent, and—above all—professionalism. A good way to do so in the television production field is with a short videotape. A demo tape is an investment in your career. You should put together a high-quality video résumé before beginning a job search.

Ron Osgood, facilities manager and lecturer in the Department of Telecommunications at Indiana University in Bloomington, surveyed over 60 professionals who either hire production personnel or who have used a video résumé to obtain their current job. The results of the survey provide valuable insights to anyone who considers submitting a videotape in support of an application for a television production position. Osgood's study sums up the opinions of those surveyed as follows:

- If you are applying for a position that relates to television production, even as a manager, you will likely be asked for a video sampler.
- Your videotape will have some degree of influence in the hiring decision.
- Your videotape should be technically very good and include a series of short clips showing a wide range of programs and styles.
- Your total tape running time should be between five and fifteen minutes in length. Longer programs probably will not be viewed in their entirety.
- Be prepared to submit your tape with your initial application. If you haven't submitted a tape and are called for an interview, bring the tape with you.
- You can use a number of tape formats, but 3/4-inch is by far the best choice.
- Let the viewer know exactly what your role was in each segment.
- Do not take credit for anything that you did not do.
- Do not put any material on the tape that you would apologize for.

In addition, Osgood developed the following guidelines for the production of a sample demo tape:

- Use the same technical and aesthetic values in your tape that you would expect to see in any high-quality program.
- Avoid using any material that may infringe copyright or cause other legal problems.
- Show only your best work.
- Use a character generator or good quality graphics to provide information on each segment of your tape.
- Consider using voice-over accompaniment to provide additional comments relating to each segment.
- If you are including material from other media formats, be sure to maintain the television aspect ratio.
- Two- or three-screen multi-image programs do not transfer well to video. Consider taping an introductory 15-to-30 seconds of the program to orient the viewer, and then modify the program to fit a single-screen concept.
- Remember that when you convert 35mm slides to videotape with the use of a film chain, you will lose approximately 20 percent of the information along the edges.
- When you convert Super-8 and 16mm films to tape, you should use projectors or special devices designed to match the frame rate of television. This will eliminate annoying flicker.

In conclusion, Osgood states, "Always consider your sample tape as a production on its own. Your creativity in piecing it together can help you achieve your goal of securing a production job."[1]

With regard to creativity, Stephen Wilson, who is writer-director of video production at Arkansas Power and Light Company, encourages producers to explore the full potential of the video

---

[1]Ron Osgood, "Guidelines for Making a Sample Videotape," *Instructional Innovator, The Journal of the Association for Educational Communications and Technology.* April 1984. p. 42.

medium in preparing a demo tape. According to Wilson, the video résumé has three basic potentials:

1. It can show what the applicant has done,
2. It can suggest what the applicant is capable of doing, and
3. It can provide some insight into the personality of the applicant.[2]

When putting together excerpts of your work, remember to alert the viewer to how well you accomplished the task. Some applicants use on-camera narration or voice-overs to introduce excerpts as well as to point out the extra production value the applicant brought to the project. When showcasing what you have done, use only the best of your work.

Most demo tapes begin and end with the first potential Wilson identified: they show what the applicant has done. Although a selection of excerpts from your best productions should be an important part of your video résumé, guard against it becoming a mere recount of your work history. Wilson's second suggestion—what you are capable of doing—is what most hiring executives really want to know.

If you are an entry-level applicant, this potential is all the more significant because you may not have a large number of programs to draw from. Use your video résumé to demonstrate your ability to meet a variety of production challenges. For example, it can include footage that you shot during your college commencement ceremony or college football game. This will demonstrate your ability to cover a live event. Footage you shot of a house collapsing under a falling tree during the worst snowstorm of the season could demonstrate both your ability to work under difficult environmental conditions and your sense for what makes news.

---

[2]Stephen M. Wilson, "Producing a Video Résumé," *International Television, The Journal of the International Television Association.* August 1983. p. 22.

It is not customary to include insights into your personality on a video résumé. But according to Wilson, "Remember, you want to use the video résumé to make it easy for the employer to hire you. It is easier to hire someone that you feel you know."

Creative video résumés run the risk of being cutesy. Be sure not to underestimate your viewer's level of sophistication. Remember, too, that your tape will not be seen under the best viewing conditions. Managers are busy people, and your tape may be viewed amid many distractions. Therefore, the segments included should be short and the explanations concise.

Be prepared to ship out or drop off a demo tape to the hiring executive. Allow for a reasonable amount of time before you get persistent about its return. It will serve you well to have several copies of your demo so that while one is being reviewed, others are available for screening somewhere else even at a moment's notice.

RÉSUMÉ AND COVER LETTER

Your written résumé must be clear and to the point. It must be free of errors, and the layout should be clean and neat. It should look organized and perfect in every respect. Remember, it is a reflection of your personality, to some extent.

A résumé should be designed to meet the specifications of the job for which you are applying. Hence, it is advisable to have more than one résumé. Better still, if you have access to a word processor, maintain your résumé on a floppy disk so that you can tailor it to the requirements of each job you are applying for.

Your résumé must be an honest account of your work history, but you need not list every job you have had. Restrict it to media-related jobs and those in which the experience would prove valuable in the position you are seeking. For example, your experience with the Emergency Medical Service will clearly give your résumé an edge when you apply for a job in a hospital or health care media department.

A résumé is your way of presenting or "selling" your skills to a potential employer. Therefore, it should reflect all pertinent information regarding your education, skills, employment experience, and career goals. You can use any of a number of different formats. Take a look at the different formats and styles presented in sample résumés found in reference books that offer detailed advice on preparing a résumé. Such books may be checked out from the library or purchased at a bookstore. Your local photocopy shop may also maintain a file of sample résumés. The format you choose should be one that will set you apart from other applicants as well as one that will be attractive enough for the hiring executive to remember you and grant you an interview.

Here are some suggestions that may be useful to you when designing your résumé:

- Your résumé should contain your name, address, telephone number, work history, education, awards and honors, professional appointments, and association memberships.
- A résumé should not exceed two typewritten pages, and it is better if it fits on one. However, be sure to leave enough margin and white space so that it does not look cluttered.
- List your work experience (if you have any) before your education and training. For most production jobs, work experience is valued more than higher education without experience.
- All items on your résumé should be listed in reverse chronological order (that is, with the most recent items first).
- Stress the link between your skills and the job responsibilities to be assumed.
- Use action words (such as planned, initiated, developed, managed) to indicate your competence. Such words also

show the prospective employer that you have handled responsibility.

Each cover letter should be personalized and targeted to the individual company executive. The cover letter should specify the job or type of work for which you are applying. Keep the cover letter as brief as possible. Remember that the cover letter is your first sales pitch, so it should be an attention-getter. It should be creative, yet professional. It should be straightforward, lucid, and informative without repeating the contents of the résumé. Use it as an opportunity to highlight the experience and training you've had that best match the job requirements. End your cover letter with a statement indicating that you will call to set up a personal interview.

JOB INTERVIEWS

The impression you make at an interview is extremely important. The two points to remember for a successful interview are:

1. Be prepared.
2. Be yourself

Bring an extra copy of your résumé to the interview just in case the hiring executive does not have it at hand.

As part of your preparation work, you should thoroughly research the company at which you are seeking employment. You should find out everything there is to know about the company—its clients, its philosophy, and its corporate "culture." Many corporations have annual reports which you can ask for and study. These reports will give you an overall idea of how the company is structured and how well it is doing.

Also find out as much as possible about the professional experience of the executive you are interviewing with. This will help you establish any grounds of common interest that can help set the tone for a warm and productive interview. Be yourself at all times—don't play a role. If you put on an act, you will find it difficult to keep it up under detailed inquiries, close scrutiny, and stressful situations.

Your dress and manner should be appropriate for the position you are seeking. A job interview is one occasion when first impressions do count. You will be judged, to some extent, on how well-groomed and styled you are.

Presentation skills are another important factor in making a good impression. If your command of the spoken language is not up to par, consider taking lessons and practicing speech. A video technician will perhaps be forgiven for stumbling over words. But for job functions that call for effective communication with other people, speech defects will not be overlooked. You should come across as a confident and self-assured person. Remember, you will not get a second chance at making a great first impression!

Hiring executives in the television industry look for good interpersonal skills in the candidates they interview. You must demonstrate the ability to listen without interrupting. A good listener observes the speaker attentively and actively participates in giving feedback. Be careful not to seem to control the conversation or to talk too much.

Be at ease as much as you can. Answer questions in a direct, brief, and accurate manner, but try not to seem abrupt. It is important to be pleasant. When you respond to a question, take the opportunity to emphasize your strengths. For example, you may be asked what kind of experience you have had with lighting sets. Instead of merely listing the situations, you can talk specifically about how you overcame the difficulties posed by multiple mirrors in a beauty salon. But don't ramble on. Be brief and focused when giving examples.

Most interviewers want to know more than what is stated in your résumé. They quickly get an idea of what you have accomplished and then need to hear what you think you can do for the company. Discuss the company's or department's needs and your abilities to help meet them. This should impress the interviewer. It will demonstrate your interest in and enthusiasm for the organization and will establish an image of you as a bright and intelligent person.

Media executives like to hire "high-energy" people. Be positive and show enthusiasm for the job. You must indicate that you have both the energy and the commitment it will take to get the job done. One executive commented that among candidates who qualified equally, he gave the job to the one who asked for it.

Before you leave, make sure you know what the next step is—will the interviewer get in touch with you or should you call?

Follow up your interview with a letter of thanks. It should be brief and include a sentence reinforcing your interest in the job.

CHAPTER 11

WOMEN AND MINORITIES

Opportunities for women and members of minority groups are increasing in some areas of the television and video industry. Although breaking into commercial television has been difficult, the percentage of women employed increased from 35.8 percent to the total number of employees in 1987 to 37.2 percent in 1991, according to the FCC Equal Employment Opportunity Trend Report. Minority employment percentages rose from 18 percent to 19.3 percent of the same five year period.

Table 1
1991 Employment at Commercial Television Stations

Job Category	Total	Female	Male	Minority
Officials & Managers	9,737	3,123	6,614	1,184
Professionals	20,095	7,755	12,340	3,467
Technicians	20,096	2,915	17,181	4,396
Sales Workers	5,810	2,822	2,988	625
Office/Clerical	8,467	7,480	987	2,490
Crafts Workers	547	156	391	139
Operatives	393	64	329	126
Laborers	134	17	117	61
Service	318	50	268	180
Total Jobs	**65,598**	**24,382**	**41,214**	**12,660**

Source: FCC BEE018-01 Pages 769–770.

At noncommercial television stations, the percentage of women employed increased from 43.6 percent in 1987 to 46 percent in 1991. The percentage of minority representation at these stations increased from 18.5 percent in 1987 to 20 percent in 1991.

Table 2
1991 Employment at Noncommercial Television Stations

Job Category	Total	Female	Male	Minority
Officials & Managers	1,810	721	1,089	209
Professionals	3,084	1,613	1,471	530
Technicians	2,374	388	1,986	526
Sales Workers	141	98	43	14
Office/Clerical	1,578	1,378	200	478
Crafts Workers	58	8	50	18
Operatives	33	4	29	14
Laborers	10	0	10	2
Service	84	11	73	55
Total Jobs	**9,172**	**4,221**	**4,951**	**1,846**

Source: FCC BEE018-01 Pages 775–776.

Representatives of commercial and noncommercial television stations state that, in general, "stations are meeting, and in several cases exceeding, EEO guidelines."

In the cable television industry, the percentage of women employed increased from 40.4 percent in 1987 to 41.8 percent in 1991. The percentage of minorities employed increased from 19.5 percent in 1987 to 24.6 percent in 1991.

Table 3
1991 Employment at Cable Television Stations

Job Category	Total	Female	Male	Minority
Officials & Managers	14,932	5,409	9,523	1,827
Professionals	4,608	2,171	2,437	723

Job Category	**Total**	**Female**	**Male**	**Minority**
Technicians	15,990	1,213	14,777	3,417
Sales Workers	11,919	5,611	6,308	3,425
Office/Clerical	33,005	28,508	4,497	10,204
Crafts Workers	9,081	525	8,556	1,824
Operatives	16,126	969	15,157	4,523
Laborers	887	100	787	259
Service	223	81	142	52
Total Jobs	**106,771**	**44,587**	**62,184**	**26,254**

Source: FCC CEE009-01 Pages 325–326.

According to representatives of cable television stations, women are making great strides in some occupational areas. For example, under the FCC classification of professionals, women in cable TV occupy 47.1 percent of the positions. In the area of sales and marketing, white women held 44.4 percent of the positions in 1987, they held 47.1 percent in 1991. While the percentage of minority employment has been increasing in the last couple of years, the pace has been slow.

CAREER PLANNING

A well-qualified and capable person need not fear discrimination on the grounds of sex or minority status. The law provides for equality of opportunity. In addition, women and minority employees who have already distinguished themselves in the television and video industry have had a positive influence on management and have paved the way for those to come. Hence, women and members of minority groups are increasingly being accepted and gaining access to jobs in various areas of the television industry. However, the competition is stiff, so women and minority groups should plan their careers carefully and be aware of the opportunities open to them as a result of their special status.

In addition to the career preparatory steps discussed throughout this book, experts offer the following suggestions:

- Get the qualifications and develop the competencies you need to work in this business. You must be both competent and qualified.
- Be willing to go where a job is available. Do not limit yourself to major markets or certain parts of the country.
- Demonstrate your talent and dependability—be willing to take on responsibility.
- Demonstrate your ability to learn—be willing to start at the bottom and do the work assigned with enthusiasm.
- Do not anticipate special consideration or discrimination.
- Utilize the assistance offered by organizations specially geared to serving the needs of women and minorities.
- Acquire broad-based experience and skills. Move from specific tasks or job areas to general administration if you want to move up into management.
- Be patient. Do not expect to move into management too soon.
- Promote yourself.
- Be ambitious and set your goals high.

ASSOCIATIONS

Several organizations offer professional assistance to women and minority groups. For example, Women in Cable is specially geared to serving the needs of women who work in the cable TV industry. The association has more than 1,500 members in 20 chapters nationwide. It sponsors a national cable management conference annually as well as several executive development seminars for women with senior management potential, year-round. The association offers a certification program in cable management.

Women in Communications, Inc. offers employees and job-seeking WICI members nationwide, a free media job-matching service. To join WICI in order to utilize the service to find a job, call (703) 528-4200.

Associations specially geared to the interests of women and minority groups include:

American Women in Radio and Television (AWRT)
1321 Connecticut Avenue N.W.
Washington, DC 20036

Association of Black Broadcasters
6860 Plainfield Road
Cincinnati, OH 45213

National Asian American Telecommunications Association
346 Ninth Street, Second Floor
San Francisco, CA 94103

National Association of Black Owned Broadcasters
1730 M Street N.W., Suite 412
Washington, DC 20036

Women in Cable
500 North Michigan Avenue, Suite 1400
Chicago, IL 60611

Women in Communications
P.O. Box 17460
Arlington, VA 22201

Women in Satellite Communications
Satellite Reception Systems
145 North Columbus Road
Athens, OH 45701

INTERNSHIPS

Scholarships, internships, and other tuition assistance programs are offered by several organizations. Women in Cable has established the Betsy Magness Fellowships for the following purposes: 1) to fund the educational endeavors of talented but underprivilcged young people interested in a career in cable TV; and 2) to provide a funding source for the professional development of women with management potential who are already working within the industry.

Some organizations that offer internships to women and members of minority groups are:

WAVE-TV
P.O. Box 32970
Louisville, KY 40232

Walter Kaitz Foundation
P.O. Box 11080
Oakland, CA 94611

Time, Inc. (Scholars Program)
Time and Life Building
541 North Fairbanks Court
Chicago, IL 60611-3333

WHMM-TV
Howard University
2600 Fourth Street N.W.
Washington, DC 20059

CHAPTER 12

FASCINATIONS AND FRUSTRATIONS

This is a true story of the aspirations and frustrations of Lynn Yeazel, western region manager, Panasonic Optical Disk Systems. Yeazel has been involved with the television industry since his high school years, which were during the early years of color television in the United States. Yeazel's career story is presented here so as to give you a feel for the reality of working in this business. His career spans three decades of working in many job areas of the television and video field—from producer-director, to innovator, to marketing manager, to entrepreneur. In the following pages, Yeazel describes in his own words his career path in light of the different areas discussed in this book: broadcast television, nonbroadcast video, the role of professional associations and networking, and the need for continuing education in professional development.

LYNN YEAZEL TELLS HIS STORY

Not unlike most people fascinated with this industry, I started on the production side of the business at a small NBC affiliate, working part-time during high school. After high school, coupled with four years of practical experience in television, I headed for the University of Wisconsin in Madison.

There really wasn't a formal training path in most colleges or universities during those days (1961). Confronted with wanting to follow a course of study that would prepare me for a career in television production—not film or film theory—I found few defined courses of action. Theatre or drama, film, speech, advertising, and engineering were some of the broad categories of study offered. The B.S. degree in those days included 120 credits of science in addition to your major.

I began to work with the few professors in the speech department who were beginning to design a curriculum for broadcast and television production. One of those professors was Jerry McNeely, who went on to write and produce for "Playhouse 90" and "Studio One." He won an award for a "Hallmark Hall of Fame" and has generally been very successful at two careers—professor and writer-producer. The other professor was Lee Dreyfus, who began exciting innovations at WHA-TV, the on-campus educational station, and later became governor of Wisconsin.

Stepping Stones

The best training and guidance during my undergraduate years really came from a part-time job at WHA-TV. With my previous local station experience, I hit the studio running and was directing live shows by the end of my freshman year. The early 1960s were exciting and innovative times for educational television. We dared to do what commercial television couldn't afford to do, as audiences would have been too small. We worked with incredibly limited budgets and learned to do a lot with a little. Perhaps, that's what made for our creative best.

In great contrast to WHA-TV, I had the opportunity to spend four summers as an intern at NBC in New York. The internship plan had three tracks: 1) production, 2) engineering, and 3) management. My partner on the engineering track helped me absorb at least an electrical engineering degree. The NBC experience was outstanding—it gave me the scope and training needed for very high pressure situations that were dollar dependent. The WHA-TV experience gave me the perspective of time, experimentation, and community service. It was during these years that I was often heard saying, "I will never sit behind a desk. I always want to be actively producing and directing." Well, I did actively produce and direct for over 10 years, but the desk did evolve.

On obtaining my bachelor's degree in radio-television communications, I stayed on at the university and got a master's degree in instructional media, again carving out a custom curriculum through the school of education. I worked in teacher education and produced educational TV for three years before exploring the newly evolving world of corporate video.

Early Days at IBM

Upon reading an article in *DataMation* magazine about IBM's internal video facilities, I wrote to their education manager in San Francisco. A long, convoluted series of letters and phone calls led me to join the Field Engineering Education Center in Rochester, Minnesota, to establish a video facility in an education center that had been classroom- and print-based.

Joining the corporate world of IBM began a slow evolution from studio to sitting behind that often-maligned desk. The Rochester Media Center was building a new facility from scratch, and that gave me the opportunity to use my production and engineering experience. I designed the facility, specified the equipment, and interviewed and hired a staff of 80 developers, artists, camera crew, and engineers. Still holding onto the love for production, I actively produced for two years and then began to do more managing and only three or four shows per year.

Building a People Network

It was about the same time when I began to become disgruntled with the National Association of Educational Broadcasters (NAEB). While they may have met the needs of their broadcast membership, they were really out of touch with members like me who produced educational video but did not broadcast the programs. In the fall of 1969, at the NAEB annual conference, I met six other corporate television people in a special interest group session on corporate facilities. Tom Hatcher from Equitable, Dick Van Deusen from Prudential, Ed Palmer from New England Telephone, Ken Winslow from Ampex, and a few others began to explore the idea of a professional association for nonbroadcast television people. The National Industrial Television Association (NITA) was born. Later, it merged with the Industrial Television Society (ITS) to form the International Television Association (ITVA).

My heavy involvement with the associations from 1969 to 1976 really exposed me to the vast opportunities that were available to someone who really always only wanted to produce and direct. The chance to work with association budgets, people, and conferences gave me the management experience I would never have been allowed to get on the job at the time—I would be regarded as too young and inexperienced to wear management boots.

National and international exposure (two trips to Europe on behalf of the ITVA) broadened my horizon as well as my awareness of what I could do, in addition to production and media center management. In the early days of corporate video, and today as well, there is a strong component of marketing and sales within your own division and within the company at large. My ITVA exposure and reputation made me more visible within my own company, opened doors, got me listened to, and generally accelerated my career.

At IBM, I began to press our education-oriented Media Center to take on the production of communications and marketing programs for other divisions of the company, most of which were in Westchester County, New York. It took time, but slowly we got to do more. It was great for my staff as the creativity of each individual was challenged. Our innovative and creative programming earned us, and the department, widespread visibility. Soon opportunities in communications, marketing, and advertising outside the company sprang up for many of us.

Behind Boardroom Doors

One of those East Coast visits exposed me to the communications group at IBM headquarters in Armonk, New York. A two-year working friendship resulted in an offer to join the secret videodisc development project called Castle. Project Castle was a step back into production, after five years of almost full-time management. I was once again faced with the job of producer, but this time using contract facilities and people. The experience is best described as production management.

Before I had time to think about it, I was project-managing two major programs, one in New York, and the other in Los Angeles. Bicoastal commuting began. The budgets, production values, and look of these programs had to be better than broadcast television. It was an ideal opportunity to bring together elements of my past work experience at NBC with educational television, and of course, the engineering background. The job was ideal. The product was not.

After three years and gallons of red ink, Project Castle began to fold in 1979. I found jobs for my staff and then began to look around within IBM for what I would do next. Prospects were bleak. Bleak because it was very hard to find anything that compared with the "ideal job" I had had. Education-training and video were still in my blood, so I began to make arrangements to leave IBM and join Walgreens Corporation in Chicago as director of education. My management at IBM began to stall me and suggested something very secretive was in the works and would be ideal for my career.

The Excitement of New Technology

Over a soggy vending machine sandwich in Poughkeepsie, I met the soon-to-be president of a new joint venture between IBM and MCA—DiscoVision Associates. There had been a heavy lid of secrecy over the formation of this company. Soon I was offered three things that seemed the most important to me at that time: 1) an ideal job; 2) a significant increase in salary; and 3) relocation to California. On a few promises and a handshake, I accepted the job and was in California 24 hours later for the press announcement.

DiscoVision provided me with the opportunity to pioneer again. From educational TV in the early days, to NBC in the early color days, to IBM in the beginnings of corporate television, I once again had a chance to work in a new industry with an exciting new technology. There were no guidelines for producing videodiscs. No technical specifications existed. There were no experts yet. From 1979 through 1982, I had what I consider to have been a once-in-a-lifetime opportunity.

I began with six team members to build the management structure of DiscoVision (DVA). We became experienced corporate real estate brokers (had to establish headquarters and find a manufacturing site); hired a new breed of video, computer, and technical program developers to help "hand hold" customers through a very complex and, at the time, unreliable process; wrote procedures and specifications that defined a new standard in a new industry; and created a new video career path with a job description for the title of program development specialist. Part of the challenge to communicate so much new information resulted in authoring many manuals, technical specifications, and marketing-oriented support manuals.

The newly evolving videodisc industry provided great career opportunities. I was able to move from an entry-level management position to senior management and a directorship in only two years. Had I stayed at IBM, it may have taken over a decade or more to make such progress up the corporate ladder (if at all).

Perils of Pioneering

But with rapid growth came some unwelcome high risk. It didn't seem risky at the time. The taste of success was heady, and everyone seemed to be revelling in it. But soon, the instability of the new business; the ill-fated partnership of IBM, MCA and Pioneer; a very complex technology; and a consumer market that was rushing to embrace tape while ignoring disc—all spelled the end of DiscoVision.

When Pioneer took over in 1982, they kept 45 out of the 450 people. Fortunately, I was one of the 45. As exciting as the corporate world can be, I witnessed its pitfalls from 1982 through 1985. The effects on people's careers and corporate profit-and-loss sheets were equally devastating. It seemed very specific to the video industry and very personal then. In retrospect, it seems to be part of the change cycle in all industries, and particularly to jobs that have evolved in the last five years. One need only look at the phone companies and airlines to feel that the video business maybe isn't quite as crazy and volatile by comparison.

Entrepreneurship

Career counselors will tell you that at some time in your working life you should take the risk of striking out on your own: "If you want your own company, do it while you're young. If you don't at least try it, you'll go through your career unfulfilled, always wondering what it would have been like." I've heard such comments frequently, and I really don't know how valid they are. I believe it is entirely possible to go through a career without being an entrepreneur, but I did try it in 1985 as executive vice president of MarketDisc, a small start-up company.

Keying off my experience in production, management, marketing, and engineering, as well as my familiarity with the videodisc industry, I was a prime candidate to tackle putting together the first sponsored point-of-purchase system for Builders Emporium's chain of do-it-yourself stores. It was an incredible experience in doing without resources.

It was also a learning experience in judging how much funding is needed to accomplish a given task in the videodisc business. Had MarketDisc been funded with a modest amount of operating capital, it would have lived up to all expectations. It did not. I left in the first year when there wasn't enough money to pay two people. Fortunately, the company has survived, and the Builders network is about to expand with new IBM InfoWindow hardware (the original hardware supplier went bankrupt).

From my brief stint as an entrepreneur, I went on to work as a consultant. This, too, was an excellent exposure to another career. Being your own boss is a real experience that cannot be appreciated until tried. Pioneering projects hit again as I set up the facilities and began the marketing plan and product offerings for my client, Color Systems Technology, a firm established to electronically colorize old black-and-white films. This company was an experience in proper funding, high technology, and prudent budgets (by Hollywood standards). The business, though under a great deal of controversy, is doing well. I completed the tasks as assigned and then went back to my old love—the videodisc business.

Intrapreneurship

In 1986, I joined 3M as a marketing analyst on their Optical Recording Project. 3M offered the warm, comfortable feeling of being part of a large corporation (IBM days revisited). It was also an opportunity to be in the classical marketing job of planning for and pricing products and services in a new technology. Once again, I sharpened my pencil to write business plans, make strategic marketing decisions, and structure and work with a field sales force. The people, products, and company were outstanding. Only one drawback for me: success in the field brought the opportunity for promotion to headquarters in St. Paul, Minnesota. But, to say nothing of my love for California, returning to the Midwest was not in the cards. Once again change was in the air, and once again contacts made while on the job brought the offer of the next job.

TEAC Corporation of America, a West Coast-based company, offered me the position of national sales manager for the Recordable Videodisc Division. Again, this job allowed me to be an "intrapreneur"—to build a dealer and systems integrator network, to conduct the associated sales training, to develop a marketing strategy and supervise publicity, advertising and other promotion efforts. It gave me the excitement of pioneering a new project, as the company was just getting into the recordable videodisc business. Having come on board at the start, I had the opportunity to interpret what the user applications may be and, in turn, be the architect of future products by communicating user needs to the factory. I had a sense of deja vu here, a feeling of doing it all over again. It was much like the DiscoVision opportunity in 1979.

When I traveled to Japan for the first round of product planning meetings, I was struck by the similarity of the challenges I had. Almost a decade had passed, and yet somehow, the diverse tasks of creating a team and bringing order and organization to a department, and of raising user awareness of a new product line and stimulating the momentum of its sales, was still exciting.

While the TEAC job was ideal at the time, the product matured too quickly and the marketplace began to change. At the close of an international tradeshow, I was approached by one of my "friendly competitors" at Panasonic. They had an opportunity in the western region for a manager of all disc products—recordable videodisc and the new (at the time) digital product line of WORM and rewritable drives. The offer was too good to refuse but with it came the task of resigning from TEAC and leaving a great group of co-workers. It was probably the worst night's sleep ever, the night before I had to tell my boss I was leaving. He had every right to escort me to the door since I was going to a competitor. He didn't! Instead he asked me to stay on as long as possible and to help train my replacement. I did stay as long as possible–leaving my office at TEAC at 4:30 p.m. to catch a flight to Las Vegas and the 1989 NAB Convention where I officially joined Panasonic that night.

The Panasonic experience has been rewarding—from the vast array of products to the large corporate infrastructure—a significant adjustment is required. As the old saying goes, "you just can't hang a quick U-turn with the Queen Mary." Once again my production experience was valuable as the Panasonic WORM and erasable videodisc products found their way into non-linear editing systems at LucasFilm, Paramount, MCA, and the general Hollywood-Burbank community. Again a bit of *deja vu*. During my third year at Panasonic, the parent company Matsushita bought MCA-Universal Studios. At a technology day briefing, I found myself presenting to many of the same people I had worked with at MCA, a decade ago during the DiscoVision venture.

There is a repeating phenomenon in this technology-based business. Initial excitement and euphoria over a new product and its marketing and promotion are followed by serious business management of production, sales, and distribution. Several model changes or improvements are instituted, and the product matures. At maturity there is a sense of sadness. New technologies or methods for accomplishing the task are seen as more cost-effective, and slowly sales fall off and the excitement wanes. It's almost like watching a child grow up and leave home. The next step? The next step is often to move on to the next product and that's where the experience at a company as large as Panasonic is so beneficial. Today, I am beginning to get involved with the new 3DO technology. It may appear that "here I go again" with a new technology that needs coordinating between hardware, software, and videoware producers. Whatever the new multimedia technology turns out to be, it won't be the classic definition of multimedia at IBM in 1969, where our media development group produced multimedia for the Field Engineering training group. Multimedia then was videotape, slides, 16mm motion picture film, audio cassettes, audio cassette-filmstrip, workbooks, overhead transparencies, mockups, and the venerable, never-changing flipcharts!

Comments

Throughout my years in the industry, ITVA was the one common thread that was never broken. The association—with its staff and members, participating at conferences, and generally being part of the "family"—kept my world in perspective. Friends, corporate associates, and the ITVA family have been extremely important to me in maintaining some sort of stability over the past five years that have seen two companies fold, and in my personal life, a divorce. That part of my life has changed, too. From socializing with co-workers at Color Systems, I met Linda, and we have just celebrated our sixth anniversary. As painful as change can sometimes be, it is so often for the *better.* I cannot overemphasize the importance of joining a professional association.

Advice to people launching on a career in the video industry:

1. *Mix your coursework.* Get some drama, public relations, advertising, writing, and production experience.
2. *Take some business courses.* I don't recommend that everyone should get an MBA, but business-related courses are *imperative.* I learned to do department budgets, business plans, etc., the hard way: on the job. A chance to experiment

in the realm of the academic environment could have saved me some costly mistakes in the real world.

3. *Get a part-time job.* If you know which industry you would like to work in, test it and try it out in a part-time job. It will give you a feel for the industry as well as whether or not you enjoy working in that field. A very positive or very negative first experience should be tempered by trying another employer.
4. *Learn to write well.* Good written communications is very important in selling ideas up the corporate ladder. I've learned to write too much, too little, and about right. It is important to know how to do each and to know when to do each.
5. *Understand statistics.* I have learned the importance of brief statistics. It is helpful to understand statistics, but wise to learn that unless you are doing a thesis, managers want net numbers—their simple, concise meaning, and recommendations based on the numbers.
6. *Read the trade press.* Keep up with your industry and allied industries by reading the many monthly "circulation-controlled" publications. These trade magazines and journals are free. Even read the ones that are not free; you may find them at your local library. For me, the trade press has been indispensable. Over the last almost thirty years, I have regularly read video- and media-related trade publications such as *AV Video, Video Systems, Videography, Videodisc Monitor, New Media,* etc. I am now starting to work with a technology I first read about only four months ago in the trade press.

APPENDIX A

PERIODICALS AND DIRECTORIES

A/V Video
Montage Publishing, Inc.
701 Westchester Avenue
White Plains, NY 10604

Broadcast Engineering (BE)
Intertec Publishing Corp.
9800 Metcalf
Overland Park, KS 66212

Broadcasting & Cable
P.O. Box 6399
Torrance, CA 90504

Cablevision Magazine
Capital Cities ABC
825 Seventh Avenue
New York, NY 10019

CD-ROM Professional
Pemberton Press Inc.
462 Danbury Road
Wilton, CT 06897

CD-ROM World
Meckler Corporation
11 Ferry Lane West
Westport, CT 06880

Cineaste
200 Park Avenue South
New York, NY 10003

Computer Pictures
Montage Publishing, Inc.
701 Westchester Avenue
White Plains, NY 10604

Desktop Video World
TechMedia Publishing Inc./IDG Co.
80 Elm Street
Peterborough, NH 03458

Film & Video
Optic Music, Inc.
8455 Beverly Blvd., Suite 508
Los Angeles, CA 90048

Imaging Magazine
Telecom Library Inc.
12 West 21st Street
New York, NY 10010

Millimeter
826 Broadway
New York, NY 10003

New Media
Hypermedia Communications Inc.
901 Mariner's Island Boulevard, Suite 365
San Mateo, CA 94404

The Producer's Masterguide
330 West 42nd Street, 16th Floor
New York, NY 10036

Teleconference
Applied Business teleCommunications
2401 Crow Canyon Road, Suite 310
San Ramon, CA 94583

TV Technology
Industrial Marketing Advisory Services Inc.
5827 Columbia Pike, Suite 310
Falls Church, VA 22041

Television Digest
Warren Publishing Inc.
475 Fifth Avenue
New York, NY 10017

T.H.E. Journal
Technological Horizons in Education
150 El Camino Real, Suite 112
Tustin, CA 92680

Videography
P.S.N. Publications
2 Park Avenue, Suite 1820
New York, NY 10016

(The)Video Register
Knowledge Industry Publications Inc.
701 Westchester Avenue
White Plains, NY 10604

Video Systems
Intertec Publishing Corp.
9800 Metcalf
Overland Park, KS 66212

Video Technology News
Phillips Business Information, Inc.
7811 Montrose Road
Potomac, MD 20854

APPENDIX B

PROFESSIONAL ASSOCIATIONS AND SOCIETIES

American Women in Radio and Television
1101 Connecticut Ave., Ste. 700
Washington, DC 20036

Association for Multi-Image International, Inc. (AMI)
8019 North Himes Avenue, Suite 401
Tampa, FL 33614

Association of Audio-Visual Technicians (AAVT)
2378 S. Broadway
Denver, CO 80210

Association of Independent TV Stations (INTV)
1300 19th Street, N.W.
Washington, DC 20036

Association of Independent Video & Filmmakers (AIVF)
625 Broadway, 9th Floor
New York, NY 10012

Association of Visual Communicators (AVC)
31942 Kings Park Ct.
Westlake Village, CA 91631

Audio Engineering Society (AES)
60 East 42nd Street
New York, NY 10065

Broadcast Education Association (BEA)
1771 N Street, N.W.
Washington, DC 20036

Directors Guild of America (DGA)
7920 Sunset Boulevard
Hollywood, CA 90046

Health Sciences Education Media Association (HEMA)
P.O. Box 771
Riverdale, GA 30274

Health Sciences Communications Association (HESCA)
6105 Lindell Boulevard
St. Louis, MO 63112

International Communications Industries Association (ICIA)
3150 Spring Street
Fairfax, VA 22031

International Radio & Television Society (IRTS)
420 Lexington Avenue
New York, NY 10170

International Teleconferencing Association (ITCA)
1150 Connecticut Avenue, N.W.; Suite 1050
Washington, DC 20036

International Teleproduction Society (ITS)
565 Fifth Avenue; Suite 416
New York, NY 10017

International Television Association (ITVA)
6311 N. O'Connor
Irving, TX 75039

Institute of Electrical & Electronics Engineers (IEEE)
345 E. 47th Street
New York, NY 10017

National Association of Broadcasters (NAB)
1771 N Street, N.W.
Washington, DC 20036

National Association of College Broadcasters (NACB)
71 George Street, 2nd Floor
Providence, RI 02912

National Association of Television Program Executives (NATPE)
10100 Santa Monica Blvd.
Los Angeles, CA 90067

National Cable Television Association (NCTA)
1724 Massachusetts Avenue, N.W.
Washington, DC 20036

National Computer Graphics Association (NCGA)
2722 Merrilee Drive
Fairfax, VA 22031

National Religious Broadcasters Association (NRBA)
P.O. Box 1926
Morristown, NJ 07960

Public Service Satellite Consortium (PSSC)
88 St. Stephen St.
Boston, MA 02115

Radio-Television News Directors Association (RTNDA)
1717 K. St. N.W., Ste. 615
Washington, DC 20006

Society of Broadcast Engineers (SBE)
P.O. Box 20450
Indianapolis, IN 46220

Society of Cable Television Engineers (SCTE)
669 Exton Commons
Exton, PA 19341

Society of Motion Picture & Television Engineers (SMPTE)
595 Hartsdale Ave.
White Plains, NY 10601

APPENDIX C

BIBLIOGRAPHY

This is a selected bibliography representing the different dimensions of careers in television and video communications.

On Job Search

Bloch, Deborah Perlmutter. *How To Write a Winning Resume,* 3rd ed. Lincolnwood, IL: VGM Career Horizons, 1993.

Bolles, Richard. *What Color Is Your Parachute?* Berkeley, CA: Ten Speed Press, 1982.

Lott, Catherine S., and Oscar C. Lott. *How To Land a Better Job.* Lincolnwood, IL: VGM Career Horizons, 1989.

On Careers in Television and Video

Ellerbee, Linda. *And So It Goes: Adventures in Television.* New York: Berkley Publishing Corp., 1987.

Ellis, Elmo I. *Opportunities in Broadcasting Careers.* Lincolnwood, IL: VGM Career Horizons, 1992.

Jurek, Ken. *Careers in Video: Getting Ahead in Professional Television.* New York: Knowledge Industry Publications, 1989.

Levinson, Richard, and William Link. *. . . !Stay Tuned! . . . An Inside Look at the Making of Prime-Time Television.* New York: St. Martin's Press, 1981.

Reed, Maxine K., and Robert M. Reed. *Career Opportunities in Television, Cable, and Video.* 2d ed. New York: Facts on File, 1986.

On Television News

Broadcast News Stylebook. New York: Associated Press, 1980.

Fang, Irving E. *Televison News, Radio News*. 4th ed. St. Paul, MN: Rada Press, 1985.

Hillard, Robert L. *Writing for Television and Radio*. Belmont, CA: Wadsworth, 1991.

On Television Production

Anderson, Gary H. *Video Editing and Post-Production: A Professional Guide*. 3d ed. New York: Knowledge Industry Publications, 1993.

Burrows, Thomas D., and Donald N. Wood. *Television Production: Disciplines and Techniques*. 5th ed. Dubuque, IA: Wm. C. Brown Co., 1992.

Ennes, Harold E. *Television Broadcasting: Equipment, Systems, and Operation Fundamentals*. 2d ed. Indianapolis, IN: Howard W. Sams & Co., 1979.

Howard, Herbert H., and Michael S. Kievman. *Radio and TV Programming*. Dubuque, IA: Iowa State University Press, 1983.

Huber, David M. *Audio Production Techniques for Video*. New York: Knowledge Industry Publications, 1987.

LeTourneau, Tom. *Lighting Techniques for Video Production*. New York: Knowledge Industry Publications, 1987.

Millerson, Gerald. *Effective TV Production*. 2d ed. Boston: Focal Press, 1983.

Wurtzel, Alan. *Television Production*. New York: McGraw Hill, 1983.

On Nonbroadcast TV

Cartwright, Steve R. *Training with Video: Designing and Producing Video Training Programs*. New York: Knowledge Industry Publications, 1986.

Marlow, Eugene. *Corporate Television Programming—Techniques and Applications*. New York: Knowledge Industry Publications, 1992.

Matrazzo, Donna. *The Corporate Scriptwriting Book.* Eugene, OR: Communicom Publishing Co., 1985.

Van Nostran, William. *The Nonbroadcast Television Writer's Handbook.* New York: Knowledge Industry Publications, 1983.

On Cable Television

Baldwin, Thomas F., and D. Stevens McVoy. *Cable Communication.* Englewood Cliffs, NJ: Prentice Hall, 1983.

Roman, James W. *Cablemania: The Cable Television Sourcebook.* Englewood Cliffs, NJ: Prentice-Hall, 1983.

Shaffer, William Drew, and Richard Wheelwright, eds. *Creating Original Programming for Cable TV.* Washington, DC: Communications Press, 1983.